Web Performance
Optimization

Web Performance Optimization

A Practical Approach

Sufyan bin Uzayr

CRC Press

Taylor & Francis Group

Boca Raton   London   New York

CRC Press is an imprint of the
Taylor & Francis Group, an **informa** business

First edition published 2022
by CRC Press
6000 Broken Sound Parkway NW, Suite 300, Boca Raton, FL 33487-2742

and by CRC Press
2 Park Square, Milton Park, Abingdon, Oxon, OX14 4RN

CRC Press is an imprint of Taylor & Francis Group, LLC

ISBN: 9781032067605 (hbk)
ISBN: 9781032067599 (pbk)
ISBN: 9781003203735 (ebk)

DOI: 10.1201/9781003203735

Typeset in Minion Pro
by KnowledgeWorks Global Ltd.

Contents

Acknowledgments

There are many people who deserve to be on this page, for this book would not have come into existence without their support. That said, some names deserve a special mention, and I am genuinely grateful to:

- My parents, for everything they have done for me.

- My siblings, for helping with things back home.

- The Parakozm team, especially Aruzhan Nuraly and Madina Karybzhanova, for offering great amounts of help and assistance during the book-writing process.

- The CRC team, especially Sean Connelly and Jessica Vega, for ensuring that this book's content, layout, formatting, and everything else remains perfect throughout.

- Reviewers of this book, for going through the manuscript and providing their insight and feedback.

- Typesetters, cover designers, printers, and everyone else, for their part in the development of this book.

- All the folks associated with Zeba Academy, either directly or indirectly, for their help and support.

- The programming community in general, and the web development community in particular, for all their hard work and efforts.

Sufyan bin Uzayr

About the Author

Sufyan bin Uzayr is a writer, coder, and entrepreneur with more than a decade of experience in the industry. He has authored several books in the past, pertaining to a diverse range of topics, ranging from History to Computers/IT.

Sufyan is the Director of Parakozm, a multinational IT company specializing in EdTech solutions. He also runs Zeba Academy, an online learning and teaching vertical with a focus on STEM fields.

Sufyan specializes in a wide variety of technologies, such as JavaScript, Dart, WordPress, Drupal, Linux, and Python. He holds multiple degrees, including ones in management, IT, literature, and political science.

Sufyan is a digital nomad, dividing his time between four countries. He has lived and taught in universities and educational institutions around the globe. Sufyan takes a keen interest in technology, politics, literature, history, and sports, and in his spare time, he enjoys teaching coding and English to young students.

Learn more at sufyanism.com.

CHAPTER 1

What Web Performance Is and Why It Matters

IN THIS CHAPTER

➤ Explaining the concept and main objectives of web performance

➤ Answering the "why it matters" question about websites

➤ Giving reasons why web monitoring is the key to success

With so much business being conducted around the world online, nowadays, the success equation of presents itself quite simply: you have to learn how to handle performance pressure and incorporate fine-tuned technologies smartly into your business. And while the pressure just keeps growing

DOI: 10.1201/9781003203735-1

1

and there is nothing much to do but to adapt and deliver, the technology part leaves plenty of room for leverage and creativity. And in today's hyper globalized and over-competitive world, there is no better way to reach and attract your potential audience/customers from anywhere in the world than through a website. Yet there is a tricky part: for the website to make a positive impact, it has to be simply outstanding.

With businesses pouring money into launching and constantly tuning the company website, it helps to know what's too much or too little to spend on professional web solutions. Thus, according to WebFX,[1] web developing and design prices range from $1000 to more than $100,000. There is a great variety of digital marketing companies that can provide you with a custom-made website. The standard design is expected to feature database integration, obligatory e-commerce functionality, and several services that could be as complex as interactive games, background music, and videos that attract attention and glue you to the site, as well as online store setup and virtual shopping cart. If you manage an enterprise, the advanced design offers a solution that matches your company's size, budget, and requests. It also maximizes your conversion rate and customers' return, which is critical for striving in a competitive market. Making sure you know exactly what features you do not need will help you to save for further investment in website maintenance that can cost from $400 to $60,000[2] and include website security and updates to improve functionality, reduce errors, and prevent viruses.

WEBSITE PERFORMANCE IS KEY TO CUSTOMER RETENTION AND ACQUISITION

It should be clear by now that delivering a performant site is vital for an ambitious business. However, there could be cases when your site may have the best products, services to offer, and unique content on the web, but if it is even slightly glitchy or boring, you could end up hurting your brand and driving potential as well as loyal customers right in the hands of the competition. Performance is a big deal, especially when it comes to matching technology with the marketing vision of your own. The website you launch therefore has to represent your spirit and make sense of what your brand, product, or service hopes to become someday. The "wrong"

---

[1] https://www.webfx.com/website-design-pricing.html, WebFX
[2] https://www.webfx.com/website-design-pricing.html, WebFX

website has to be viewed as a liability and dealt with accordingly. The "right" website will manage to transmit a clear sense of purpose. Keeping the vision and standards of operation on a high point results in capturing bigger market shares, increased efficiency, and profit. On the other hand, letting go of the site's performance issues and delays will cost you the very precious audience to the point where one day your website might be shut with its doors bolted permanently.

WEBSITE PERFORMANCE IS THE OBJECTIVE

In the early days, it was enough for a company to just draft a website that vaguely resembled a well-done online catalog. Sites were built with the aim to generate more views and consequently more sales. They were measured using quantitative tools like Google Analytics that launched on the November 14, 2005. At that early stage, website owners were trying to figure out how well their sites were performing by adding adequately ancient "visitors' counters" at the bottom of the page. Those times have surely passed, but nowadays website owners still struggle with website performance measurement. In recent years, an alternative evaluation to judge each new website as the objective on its own, looking at the site's quality and interactive performance, was introduced.

Web performance objective has shifted from calculating viewers to make sure the website is fast, offers reassuring feedback, has smooth animations, and provides continuous, strategically an important service to the business.

WEB PERFORMANCE IS USER EXPERIENCE

Fast page load time equals more visitors willing to return and more people trusting the website. Ideally, viewers expect pages to load in 2 seconds, and after 3 seconds, up to 40% of users will abandon your site.[3] Similar results have been presented by major sites like Amazon, which found that 100 milliseconds of additional page load time decreased sales by 1%,[4] and also Google, who lost 20% of revenue[5] due to half a second increase in webpage load time. Akamai has also summarized that 75% of online shoppers who

---

[3] http://www.mcrinc.com/Documents/Newsletters/201110_why_web_performance_matters.pdf, Gomez

[4] https://www.forbes.com/sites/oreillymedia/2014/01/16/web-performance-is-user-experience/?sh=1bd5c0c45a52, Forbes

[5] https://perspectives.mvdirona.com/2009/10/the-cost of latency/, Perspectives

experience page freezing or checkout process that takes too long will not buy from that site.[6]

These numbers are crucially important for understanding that great user experiences can only exist if a thorough, detailed approach to website performance is implemented. Putting performance first will save you and your customers' precious time and money. Considerations about web presentation have to be at the core of business conversations, and investing in designing sites with impressive, rich content that includes complex animations and dynamic graphics can no longer be perceived as a luxury.

Think about your most recent website design you have seen. How many font weights have you noticed? Were the images well placed? Did the overall structure and design capture your attention? When you build your own site, sometimes you will have to make choices that favor art-conscious esthetics; other times, you will opt for lesser page load time. No matter what you do, the key is to always be thinking from the user perspective, making sure you keep in mind the following major areas of web performance application.

Reducing Website Load Time

In its simplest terms, page load time is the average amount of time it takes for a page to show up on your screen. It's calculated from initiation (when you click on a page link or type in a Web address) to completion (when the page is fully loaded in the browser).[7] This includes all content on the page, such as text, images, and videos. The speed is mostly affected by latency (the time it takes from when a request is made to the time it takes for the response to get back)[8] and the count and size of the initial files. A general strategy when creating a website is to make your files as small as possible and reduce the number of Hypertext Transfer Protocol requests (HTTP—used to structure requests and responses over the Internet).[9] Additionally, with Google's Shift to mobile-first indexing, it became just as important to focus on your website's mobile loading speeds as well.

---

[6] https://www.akamai.com/us/en/multimedia/documents/report/akamai-site-abandonment-final-report.pdf, Akamai

[7] https://www.bigcommerce.com/ecommerce-answers/what-page-load-time-and-why-it-is-important/, Big-commerce

[8] https://developer.mozilla.org/en-US/docs/Web/Performance/Understanding_latency, MDN Web Docs

[9] https://www.codecademy.com/articles/http-requests, CodeAcademy

In 2019, a team at Portent,[10] a web digital marketing company in Seattle, completed internal research to find out the extent to which site speed impacts the user experience and potential earning (conversion) of any given e-commerce website. They found out that the highest e-commerce conversion rate occurs between 0 and 2 seconds with an average of 8.11% conversion rate at less than 1 second; at a 5-second load time, the e-commerce conversion rate gets down to 2.20%. After that, expected returns are largely below 2%.

They have demonstrated how if 100 people visit your website for a $50 product at the given speed, it could easily make a difference in your earnings:

- A less than 1-second page load time at an 8.11% conversion rate would result in $405.50

- A 1-second page load time at a 6.32% conversion rate = $316.00

- A 2-second page load time at a 4.64% conversion rate = $232.00

- A 3-second page load time at a 2.93% conversion rate = $146.50

In 2006, Google conducted a similar experiment.[11] It occurred that Google received requests from viewers to receive more than ten results on their search page, precisely 30 results per page. Unfortunately, doing that resulted in a change in page load time from 4 seconds to 9 seconds a page. The addition of 500 milliseconds to the search results time eventually caused a reduction of 20% in page views and ad revenue. In the same year, Amazon followed that strategy, and surprisingly, only 100 milliseconds delay in their case negatively impacted sales.

The following proves once again, in simplest terms, how site speed could affect your conversion rates and, if ignored, may lead to drastic gaps in profit.

Making the Site User-Friendly

Usability basically means user-centered design. Both the design and development process are focused on the prospective user—making sure their goals, visions, and requirements are met.

---

[10] https://www.portent.com/blog/analytics/research-site-speed-hurting-everyones-revenue.htm, Portent

[11] http://glinden.blogspot.com/2006/11/marissa-mayer-at-web-20.html, Blogpost

The very first principle of usability would then be "orderly access to your website." If your website does not work, it would automatically be regarded as useless, leaving any potential customers unsatisfied and frustrated. To avoid that, you need to invest in two basic things. First, do not economize on web hosting. A good web provider has to be dependable and ensure fully available server uptime. And, second, you need to spare time to check and fix any broken links that might exist on your website.

Nowadays, it has become vital to "make sure your site is correctly optimized and adjusted to different screen sizes: desktop, tablet version, and mobile screen." The device change therefore should not, in any case, affect the quality, clarity, and speed of the site. The outline has to be set automatically and present the original, explicit version of the website, without any clutter in content.

Visitors search for your website for a certain purpose. And ultimately, it is your job to deliver what they came for and please their initial needs. However, if their experience of your site does not match their requests or will be perceived as not user-friendly, they might not ever come back. Therefore, you should "try and connect to your audience's mental vision" of how they would wish you to form and layout the content of your site.

In most cases, in order to ease users' perception, they would require "consistency in the web design," certain predictability of the entire process of surfing the net. With that, it is also necessary to "guide them through your site," making their exploration of what you can offer smooth and seamless. When planning the website architecture, it is highly recommended to think through how you are going to order pages, what information you want to highlight on the main and individual pages. People do not usually read much into websites, rather just scan it through. So, any valuable text you wish to add "has to be accessible and readable to users"; in such instance, breaking content into body text, headings, and subheadings across the website would minimize readers' time to search what they initially came for. Apart from the written material, features such as buttons, pop-ups, scrolling, and animation have to be authentic as well as fail-free.

And at last, it is fundamental to provide a platform for "effective feedback interaction." It is an option that may not be used by your customers as often, but it is considered to be a mandatory requisite in website development.

Measuring Website Performance

Simply put, investing in website creation but not measuring its regular performance could easily be considered a waste of resources. Through site analytics and measurements, you can learn what content you might need to change and what functions you should hide or promote; you may develop digital strategies and set Key Performance Indexes for your team based on results of website performance alone. Ignoring such potential may lead to unsustainable business optimization and other faulty business regulations.

Web performance includes monitoring the actual and perceived speeds and involves a number of various metrics and tools to measure those metrics. The actual website measuring mostly deals with the number of resource requests, site size, latency, and JavaScript performance. With monitoring, you can find out and decrease any issues with asynchronous application communication, loading time, or other tools that overall interact and spoil users' experience.

Perceived performance, nonetheless, strictly manages how the website is perceived. It does so through engaging with users or by adopting their viewpoint on the application, not via quantitative and systematic metrics. Further in-depth concepts would be discussed in Chapter 3 "Performance and Speed" and Chapter 5 "Measurement and analysis."

To summarize, every website needs to have data to reflect upon and indicate how well or not the business behind the webpage is doing. The metrics system, monitoring application, and approaches vary depending on directly and clearly assigned objectives. Yet, more importantly, web measurement could be a great source of new ideas, it could push to eliminate certain patterns of inspiration to try the latest tech experiments. After all, it is a matter related to business, and that means you need to learn how to rinse and start over again when time dictates to.

WHY WEB DEVELOPMENT MATTERS?

We are proven to be absolute creatures of convenience. And one of our biggest convenience-driven achievements has been creating and accessing every day a whole new digital world that contains nearly all the necessary information about everything by simply pressing a button or clicking an icon. Assuming you are producing a product or offering services, even if those are provided in the context of a specific geographical state, you still have to make sure you are present in the World Wide Web to reach and affect the bigger audience.

It is true, however, that recently the way people search for things has changed. More people prefer social media and mobile applications as the main sources of information input and output. As a result, business owners everywhere have been forced to shift to another digital dimension and stay alert, flexible to changes to keep their audience. With that, we might start questioning whether web development still remains important or maybe having an account on Instagram or Facebook is good enough for now? The outlined seven reasons below will attempt to convince you that website presence is still as fundamental as ever.

Better Website Performance Improves Brand Awareness

All of us are digital surfers. And if you are a brand manager, your main task would be to make sure as many people as possible drift and land on your website. Your brand popularization and awareness start there only. However, as the Internet grows, search engines are becoming more sophisticated and hard to crack. Search engine optimization (SEO) has changed the dynamics of web search, shifting the focus from brands to viewers. Algorithms are tailored in the way that without a sound local and international SEO strategy, it would be hard for anyone to compete for the top spot of the search page and eventually wider brand popularity.

The journey to that goal starts from analyzing your page performance data first. Any changes that would follow up from that report will be predetermined to improve original performance indexes and factors.

It would also help to know how Google works when it assigns your site with a page and positing in the search engine. The algorithm ranks higher websites with faster load time and fewer instances of pogo-sticking. Pogo-sticking is described as a process when a Google searcher bounces from your page back to the search results, probably looking to click on another page.[12] When a viewer just glances over your website and exits back to the search page, the Google engine would presume that your content was not good enough, it failed in a way to satisfy users' requests for specific information. Hence, the higher your pogo-sticking number, the lower you drop in the search ranking. Evidently, in order to promote your brand, you need to focus on how to win users over and minimize if not eliminate your site's bounce rate completely.

---

[12] https://backlinko.com/hub/seo/pogosticking, Backlinko

To focus on the content, Google might as well disregard websites that have irrelevant or unsuited keywords in them. Therefore, for improved SEO coverage, you might need to restructure or work extra on your brand description in terms of its recorded objectives and a vision that you put out there for people to consume.

Better Website Performance Leads to Higher Conversion Rates

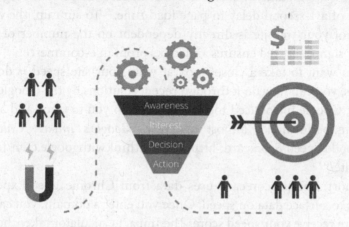

Running a business requires being in peace with business metrics in your head (most of the time). Owning a company website comes with ambitions to capture a bigger, wider, better audience at all times. Because only loyal and recently "hunted" customers can secure your sales. Thus, you have to make them happy while they are on their online shopping run. Website layout, quality, and, most importantly, the speed are crucial factors that can contribute to higher consumer satisfaction and therefore better conversion rate. We have previously talked about the importance of reducing low time, but just to stress on this once again, a research by Aberdeen, "leading global provider of behavioral-based solutions"[13] as they present themselves found that for every second above the 2-second mark, sales rates would drop by 7% and page views by 11%.

And for e-commerce, even for basic commerce, a 7% drop could be a big hit. To illustrate, imagine you own a medium-sized business website that provides sales for $10,000.00 a day, a drop by 7% means a loss of over $2.5 million per year and all because of the site speed that takes 3 seconds

---

[13] https://www.aberdeen.com/research/#whyaberdeenresearch, Aberdeen

instead of 2 seconds.[14] In conjunction with that, 11% drop in page views would cause significant damage to news providers. For example, a news site as big as Cable News Network (CNN) or British Broadcasting Corporation (BBC) could get 60 million viewers per month, with an average of four pages per visitor; 11% drop in page views means 26,400,000 fewer views a month, and with the proximity that your site made $1.00 per 1000 page views, you are looking at $31,680.00 loss in revenue per year. Again, all because of a 1-second delay in page load time.[15] To sum up, the desired success of your website is directly dependent on the number of views, sales, or signups; speed ensures and locks profit in e-commerce.

If you want to take a closer look at what your site speed is doing to business, you can now do it through recent solutions by the Google team. In 2018, Google introduced tools that can help you to track and analyze your site performance almost instantly—*Google's Impact Calculator* and Mobile Speed Scorecard: https://www.thinkwithgoogle.com/feature/testmysite/

In short, Speed Scorecard uses data from Chrome User Experience Reports to surface data on speed. Once you enter a domain, you can then and there receive your speed score. The Impact Calculator takes the speed index and other codes for monthly visitors, sales rate, and product value and then displays the approximate evaluation of potential revenue raising once the site speed is enhanced.

Ultimately, as a marketer, you would also want to look through the overall marketing campaign and make improvements where possible. For instance, not only the speed of service but also the quality of it would affect sales. It is always a good idea to route buyer's experience through the site from entering to purchase and implement other alterations when necessary. For instance, showcasing "calls to action button" on every page of the website to enable instant, open conversation and feedback from customers or installing automatic form fills that should be limited in number (because no one has enough patience for that) but saving customer information for other potential purchases.

The above described are still just basic systemic approaches. Shall you refer to a professional web developer to create a website, you would be able to optimize every page in detail to maximize usability and future

---

[14] https://blog.uptrends.com/web-performance/website-performance-matters/, Uptrends
[15] https://blog.uptrends.com/web-performance/website-performance-matters/, Uptrends

conversions. In Chapter 4, we will focus on "Conversion rates: localization and SEO-specific concerns, and on-page optimization."

Web Performance Affects User Experience

We have already discussed how fine-tuned, fast web performance matters when delivering a good (aim for impressive) user experience. It is the number of efforts you put into every little detail that would make the overall, cumulative difference in the end. Just think about it: Kissmetrics[16] data states that 44% of online shoppers once had a bad experience on the website and are going to share that warning information with their friends. And 79% of users if dissatisfied with a website quality and user-friendliness are less likely to buy from that site ever again.

On the bright side, this data should not intimidate anyone. It has to be understood that making a good website, the one that brings benefits, is not one person's work but rather a solid team effort. Challenge your marketing team to deliver better content and challenge your development team to come up with solutions that can optimize site performance and minimize fallouts. Only in collaboration and constant feedback-monitoring a welcoming environment, you would be able to deliver faster and more effective results. When thinking about your user experience, focus point could be turned to analyzing the nature of your business. Brainstorming about potential requirements, frustrations, and anticipations of your customers when visiting the website can as well help to craft a better encounter. Here are just several focal points you might consider along the way:

- Are you respecting customers' privacy and personal information safe? Is it noted somewhere in the layout?

- Is there an established interaction between customers and your brand? Are you providing them with contact information and instant feedback options?

- Do you think your website leaves room for a language barrier? What additional language options you consider your audience might need?

---

[16] https://www.monitis.com/blog/top-10-things-you-need-to-know-about-website-performance/, Monitis

- Is your website lightweight in both content and loading? Keep in mind that viewers do not want to put much effort into reading instructions neither they have the patience for it.

- Is there a sense of direction and progression when you explore the website?

In short, to make the user experience pleasant for customers as well as desirably profitable for you, it is highly recommended to take into account your main audience's desires and potential concerns when starting or maintaining website performance.

Poor Website Performance Destroys Brands

Some might argue that the worse thing than not having a website is owning one that is simply awful. But how would one describe an awful website? To give you a hint: it is the one that would make your viewers think that you have been out of business for quite some time now. Not dealing with this might cost you more than sales, it would completely devalue our brand reputation. There is a great connection between website representation and business market value. And when inadequate digital marketing disables positive customer experience, the whole company behind the brand suffers.

A successful website is a whole package deal. It has to involve features that we have mentioned before like search engine marketing strategy, has to have quick loading time, and have approachable, user-friendly interface. Every element of web performance matters. You cannot afford to let one go off if you want to have a striving business. Imagine you run an online fashion retail company. You have done everything right—invested in both developing and marketing experts. However, shall you ignore the need to design your website up to date with another online fashion store, it is going to look unappealing to the main audience and look bad on mobile screen inevitably leading to all your previous investments having very little return.

It is as simple as that establishing a sound and fully equipped online presence will affect your offline operations directly. And customer' service should matter on digital sites to the same extent as they do in real life. Therefore, staying on track with the latest solutions and e-commerce innovations is incredibly important nowadays.

To illustrate with an example, in 2013 Radware Blog[17] conducted a similar study that researched how the speed of service affects customers' brand perception. They had participants engaged with standard tasks on four e-commerce and travel websites. Researchers provided half of the participants with a full-speed Wi-Fi connection, while the other half had a distorted, poor connection that reduced page load speed by 500 milliseconds. The results were surprising. Users with delayed connection had a 26% increase in peak frustration and an 8% lower engagement with the site content. Moreover, the same group later described the website brand as "boring," "childlike," and "tacky." In contrast, the group with fast WIFI had a positive experience and reported perceiving the brand as overall friendly and the website useful.

Better Performing Websites Gain a Competitive Edge

When you observe users bouncing, or using the term that we have explained earlier, "pogo-sticking," they do not just stop their search but head straight to your competitors who might win both the sale deal and their loyalty. To avoid that and keep your brand image unharmed, you need an engaging website and a certain competitive advantage in comparison with your closest competition.

According to the Weber–Fechner Law or Just Noticeable Difference[18] for people to notice the difference between two similar things, it requires a minimum difference of 20%. So, to ensure that your audience is aware of how much better your website is, all you might need is to gain at least 20% leverage over other companies. You can choose to be faster, brighter, bolder, or on the contrary, slightly more sophisticated to highlight your unlikeness compared to other site performances. For instance, if your web animation takes 3 seconds and your competition does so in 2 seconds, your website needs to be displayed fully in 1.4 seconds for customers to see the difference, acknowledge your brand as more advanced, and choose to stick with you.

As mentioned earlier, there are several Google online solutions that can help you gain analytics about your website. Thus, you can get access to valuable and valid data that can further guide improvements in your SEO,

---

[17] https://blog.radware.com/?s=Mobile+Web+Stress%3A+The+impact+of+network+speed+on+emotional+engagement+and+brand+perception, Radware
[18] https://en.wikipedia.org/wiki/Just-noticeable_difference, last edited November 17, 2021

marketing, or web maintenance strategies. Yet such data should not be examined on its own. The best effect could come from pairing quantitative reports along with competitive analysis. Competitive analytics has the capacity to place your onsite analytics into context and give you ideas for potential marketing campaigns.[19] Conducting comparative inspection or survey will help you gain the upper-hand advantage that every business and every digital source is looking for. The study will help you to do the following:

- **Get a more profound insight into the marketplace:** Good marketing practice includes being careful about how the word "insight" is used, being especially attentive not to interchange it with raw data, quantitative research, or general knowledge. Insightful marketing should always drive better understanding of current and future trends in the market, backup informed decisions when choosing a specific expansion strategy, and identify new, innovative market options. In the e-commerce space, it all boils down to traffic. Who gets the most views and why? What are their most successful channels of interaction? What keywords do they refer to attract organic traffic? The competitive review will help you to identify what schemes and approaches other websites implement to become and stay leaders at their niche. For example, key market insights can help you to:

  - Find new revenue sources

  - Ensure website is reaching the precise number of customers for which it was aimed

  - Develop marketing campaigns and SEO strategies

  - Measure current performance against overall potential

  - Monitor brand awareness and customer satisfaction

  - Learn how your target audience actually uses your product on a daily basis as well as responds to their feedback

---

[19] https://blog.alexa.com/why-comparing-your-site-to-the-competition-matters/, Blog

- Understand what draws viewers to drift to your competitors

- Utilize previous browsing patterns or behaviors to offer customized interactions and encourage specific action.

- **Stay up to date with current trends:** The urgency of technological innovations makes quick market updates a necessity. Nowadays, automated digital market research has the ability to produce current global market insights to be produced in real time for a fraction of the time and money. When it comes to the ever-changing and evolving field of innovation, insights need to be quick.

  Being alert and aware of any slight industry changes also demands that you quickly respond to them. Preferably, at least 20% faster than our competition. Performance comparison also demands that you review not only just one, or a couple of other portals, but also the overall external situation they operate in. For instance, you could not guess why your rival site's recent traffic index suddenly skyrocketed or plummeted, when in reality it had nothing to do with their campaign but rather uncontrollable externalities that impacted the case. You would want to know that as soon as possible because it could affect your operations as well. Real-time responses to consequences that your competitors battling with (or taking advantage of) might one day save your business.

- **Benchmarking:** Let us accept that in the absence of comparison benchmarks, it would be quite impossible to know whether we ever pay enough attention or invest enough resources for business development. A competitive survey thus can assist you with recognizing which brand could be sufficiently used for your benchmark measurements and where in the marketplace there is room for your improvement. For example, if you are only present on the website and do not locate your services using social media, having found out that your competition gets most of its traffic from social media applications will force you to grow and spread on new platforms. Even if you are the leader of the industry, it should be useful to know how other progress notes that data and remains on top of things.

  The very simple and independent research that you can do is to browse other brands' websites to see how they portray themselves and describe their services. This may allow you to reflect upon how you conduct your business online and what methods use to reach more viewers. And while browsing, it is helpful to pay extra attention to customers' testimonials and success stories on your competition websites where they emphasize how helpful or not helpful other businesses were. Not only can you see how and where your competitors are regularly failing—shall it be poor customer service, unsatisfactory product quality, or something as simple as shipping delays, but you can also spot your audience or untried alternative markets.

- **Refine your style:** There are plenty of sites out there. Beyond being known for great content, it is important to make your website stand out by providing an exceptional user experience. Make sure you use tools to put yourself in your customer's shoes and then identify any site characteristics that may compromise their ability to move easily through your site and fully experience it. In turn, these optimizations can produce a positive impact on the performance of marketing campaigns.

    It is not only the website you should be looking at, but also if possible, competitors' social media pages and YouTube video content—everything that can assist in improving and refining your own sense of style and place in the digital market. You could as well analyze whether you and other brands match in your vision on an "ideal user" and use the chance to see how you can work to set apart and create your distinguished customer communication manner.

    Using that information can substantially influence the way you approach your own marketing. Exploiting others' downsides might sound cruel, but it can help to highlight your stronger points and cut on flawed functions. If you are a smaller enterprise, you can readjust and upgrade your offerings faster than bigger companies can afford. Having an advantage in time should be used accordingly and create a similar advantage in competition.

Website Monitoring Is Essential

Website development matters because it gives a chance to use website monitoring tools in business development. There are several direct benefits from investing in site monitoring—it usually increases general site performance, returns costs, and improves marketing efficiency. But most importantly, it points at exactly the area that you need to build on to let your business grow further.

Web analysis can significantly affect the way you identify with your customers. For example, you would want to know when and where they're visiting from, what gadgets they're using, what are their online requests, and another socioeconomic factor like age or gender. Acquiring these bits of information will help to better comprehend what's essential for your viewers and how to better customize their experience. However, if you consider outsourcing monitoring your website to another company, it is recommended to hand it over to an established industry leader and innovator without the intention to shortcut or shortchange the monitoring assignment.

To help you get the idea of what elements different types of monitoring metrics can consist of (in this case, the main website performance metrics, SEO performance metrics, and business performance metrics), several examples are included below. It is important, however, to emphasize that the list we offer is not entirely complete. Actually, you should be able to order and construct the monitoring methods and indicators for yourself, keeping in mind what measures would be most useful for your business.

- **Website performance metrics:** This method helps to estimate your website content reach and measure engagement rate with visitors. There are several indicators that are being reviewed for such analysis:

    - **Page views:** simply the number of times when your page is being loaded. Usually, experts are looking at some specific timeframe within which they want to check the total number of entries and potential consumption figures.

    - **Unique users:** Wikipedia gives the following description to this term: "Unique visitors refers to the number of distinct individuals requesting pages from the website during a given period, regardless of how often they visit. Visits refer to the number of times a site is visited, no matter how many visitors make up those visits."[20]

    - **Bounce rate:** is a term we had touched upon previously that defines the case when your website was briefly visited by someone who then decided to either go back to the search page, proceed to another source, or close the browser altogether. In other

---

[20] https://en.wikipedia.org/wiki/Unique_user#:~:text=A%20Unique%20User%20metric%20is,dimension)%20of%20a%20particular%20week., last edited July 29, 2020

words, it is the number of the only main page or single-page visits. Therefore, if you aim to have an engaging, interesting website, you need to monitor your bounce rate. The lower it is, the closer you are to your goal.

- **SEO performance metrics:** SEO metrics give the real-time measure of web performance in regard to how much organic traffic it attracts.

  - **Keyword ranking:** provides your calculated search engine rankings for those keywords you have targeted. It makes it possible to see your ranking dynamics over time as well. This metrics is mainly used in KPI settings as it demonstrates the effectiveness of your SEO team to attract unique and loyal visitors. You can achieve a higher ranking by identifying keywords or phrases that your certain customers will be looking for, and that would inevitably lead them to your website. Additionally, there is a number of online tools that can help you to measure your keyword ranking for free:

    - **Moz's Keyword Explorer:** https://moz.com/explorer

    - **Google Ranking:** https://smallseotools.com/keyword-position/

    - **Ahrefs Rank Checker:** https://smallseotools.com/keyword-position/

  - **Backlinks:** is simply a link from one website to another. The process of getting those links, or another word "link building," could be quite a consuming process as it represents complete confidence in another source's content and purpose of conduct. Backlinks could be used as a great tool in comparative research; in other words, once you can see your competitor's backlink ranking and the list of sites that vouch for their confidence and partner with him, you can discover what partnership or cooperation strategies may help them to get certain advantages over others. Online backlink tools like Moz Link Explorer can show how many backlinks you have.

    Homepage—https://moz.com/link-explorer

- **Business performance metrics:** Business metrics normally used to show the process and the progress of your website delivering revenue.

  - **Leads:** are the active website users that personally engage or share their information with you (e-mail, phone number). These are the premium customers, the most important people for your enterprise, as they usually the ones that convert their participation into your profit. To explain plainly, the more leads you can capture and keep over time, the more income you get for the company.

  - **Conversion rate:** should not be confused with conversion count. Conversion rate is the percentage of website users that took the ultimate desired action. Usually, users tend to limit that only to e-commerce services. However, the desired action does not necessarily mean purchase, and it could be a review or subscription. Shall you have difficulties in calculating conversion rates on your own, you can always rely on a number of online tools like WebFX that could be very useful when drafting yet another business plan.

  - **WebFX Conversion Calculator:** https://www.webfx.com/tools/conversion-rate-calculator/

When shopping for monitoring solutions, you need to look for one that provides instant alerts to be able to fix problems and performance bottlenecks before they impact your end users.

- **You want to get alerts through preferred channels:** via email, SMS, or on your preferred notification channel using various app integrations. You should be able to select how quickly you want to be alerted and also set alert-repeats for getting notified when an issue is fixed.

- **Alert the right people:** you want to choose individuals or teams to send alerts to those based on specifics of the task, so the right people in your organization are notified and can fix issues before a user can spot them.

You want a monitoring solution to provide you with detailed reports on a daily, weekly, and monthly basis. There is a number of reports you should be sharing with your key stakeholders:

- **Uptime reports:** list uptime, downtime, and response time data for various time periods specifying every request made by servers and the location of customer's requests

- **Transaction reports:** give measures specific to each transaction

- **Page speed reports:** provide an overview of tests made and data about web performance such as load time, number of requests, page size, and others

- **Real user monitoring reports:** provide insight into user behavior patterns and analyze time patterns and specific to geographic locations, browsers, and devices

Simplified technical and nontechnical monitoring of your web performance could serve as a great tool for the following specialists in your team:

- **Digital marketers:** monitoring will help to secure site availability, gain insight into user behavior metrics, and review key transactions necessary to your marketing review.

- **For web developers:** analysis will help web developers to understand why performance issues are occurring and troubleshoot issues in a more efficient manner.

- **For IT/Web Ops:** monitoring might give insights into whether an issue is specific to a location, device, or browse as well as help to prioritize fixes that take time and effort.

Good monitoring tool should also be able to deliver reports showing the long-term impacts realized through your optimization efforts and have a capability to share those successes online with key stakeholders.

Mobile-First Design

A mobile-first design is a concept from the developer community that dictates to prioritize mobile screen size and design content for mobile devices first, only for desktop and tablets second. If you own a website or thinking about developing one, it is safe to say that you might consider starting from the mobile version.

There are two approaches that developers use to create a responsive mobile design. In order to make web or application interface display all the necessary information, at first, designers provide customized versions of product for different ends.

Progressive Advancement is the process of starting a product design with a version for the relatively lower browser. At first, this version includes the most basic functions and items. After that, in order to advance it for a mobile or a tablet, various interactions and more complicated features were being added.

Graceful Degradation, on the other hand, starts with an advanced product design that is then made more compatible by cutting some functions or contents.

As early as 2010, Google was one of the first ones to pitch the mobile-first philosophy in product design. Later, in March 2018, the company has announced that they were going to start with mobile-first indexing. In March 2020, Google issued a statement that it would start mobile-first indexing for the whole web.[21] Now, what does it precisely mean? It means that from now on, every mobile version of every site will be indexed and ranked accordingly. It might sound slightly intimidating, yet if you are certain that both desktop and mobile versions of your site have the same structured content and quality, you have nothing to worry about. The main reason why the mobile-first approach is spreading so fast is the over-whelming speed of mobile commerce (m-commerce) development.

M-commerce is defined as "the use of wireless handheld devices like cellphones and tablets to conduct commercial transactions online, including the purchase and sale of products, online banking, and paying bills."[22] Research company, Statista, estimated mobile commerce sales transactions in the United States alone to total $123 billion in 2016 and $207.2 billion in 2017. And these same numbers are growing exponentially around the world. Another reason why implementing mobile-first tactics is crucial to success is that staying unoptimized will affect your brand reputation as unresponsive, unalert, and repel the majority of mobile customers. For instance, back in 2012, Google proved that mobile-friendliness was indeed one of the main factors when making purchase decisions. Thus, 67% of respondents indicated that a mobile-friendly website made them more likely to buy a product; whereas 61% specified that a bad mobile experience made them leave website.[23]

In short, the "mobile-first" principle has an important role in web development. On the one hand, it helps to save product visibility and improve your brand. On the other hand, it forces developers to pay more attention to the content and tuning of the site in order to be able to create neat and practical mobile interfaces. However, as smartphones become more powerful, the mobile-end principle might no longer be considered as an additional practice in the near future but rather an absolute necessity that cannot be ignored.

---

[21] https://developers.google.com/search/blog/2020/03/announcing-mobile-first-indexing-for, Google

[22] https://www.investopedia.com/terms/m/mobile-commerce.asp, Investopedia

[23] http://googlemobileads.blogspot.com/2012/09/mobile-friendly-sites-turn-visitors.html, Google

CONCLUSION

Improving your site performance starts once you launch it. Web performance has to match your businesses' marketing and operation visions. Committing to website performance in terms of user experience and SEO requires consistency and systemic approach. As we have discussed in this chapter, one of the main objectives of web performance is to invest in better, optimized web as well as monitoring services. Maintaining a fast, high-quality customer experience should involve a great deal of engagement and will eventually impact the reputation of your brand. The next chapter will talk about optimization techniques in detail, covering front end, content, and back end.

CHAPTER 2

Optimization Techniques for Front End, Content, and Back End

IN THIS CHAPTER

➤ Explaining why web optimization is important

➤ Learning back-end and front-end optimization

➤ Mastering content optimization

Now that we know the importance of web development, it is time to discuss how to use it to make your website strive and reach its fullest potential. By mastering website optimization, you can expect to put your product on the map and get as many customers as you would like. The process of optimizing or improving your web performance through back end, front end, and content tuning guarantees to enable heavier, more organic traffic and higher profit.

It is crucial to position yourself or your enterprise correctly in the industry and in the digital realm. And if you do not invest in optimization, your efforts to run a successful business may become irrelevant as your site will not be noticed, ranked, and recommended to viewers. So the goal here is to make your site visible to search engines and attract targeted

DOI: 10.1201/9781003203735-2

traffic from potential customers. But where do you start optimization from? The first step would be trying to single out and recognize issues that your site is currently facing. Brainstorming with your team seems to be one of the most efficient ways to spot areas that need improving. Is it your web design? Or maybe some would find the speed too slow? Is the content catchy enough for your ideal user? Asking questions as such would form a great foundation for further actions. The core web optimization consists of back end, front end, and content development. And we would go through each one of them in this chapter.

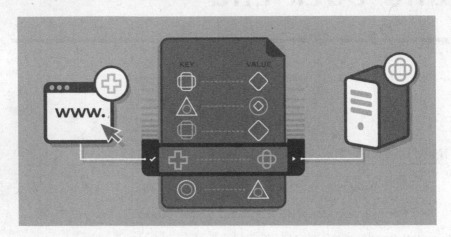

BACK-END OPTIMIZATION

Technical off-page optimization could be described as "a roadmap for search engine bots" that allows search engines to understand the intentions of your website.[1] Ideally, the back end should be crafted in a way to be able to notify bots and search engines in regards to the site's items, categories, and priorities. Off-page functions and ramifications are not seen by web views but still have great significance. You can look at it this way: technical optimization is the foundation for web operation, and once your back end is off it does not matter how your front end or web content would

---

[1] https://www.ecreativeworks.com/blog/what-is-the-difference-between-front-end-and-back-end-optimization#:~:text=back%2Dend%20optimization%3F-,Back%2Dend%20Optimization,engines%2C%20crawlers%2C%20and%20bots., Creative

look like anymore. There is a number of standard modules that you may concentrate on to raise off-page efficiency.

Check for Any Sitemap Errors

Creating a sitemap has never been easier. However, even if you are using an online sitemap generator, you may make a mistake with formatting and submission. Usually, first-time developers stumble upon compression issues when Google refuses to process your sitemap because it could not be uncompressed. It could also be that you are having some HTTP glitch when downloading the map or something as simple as submitting the sitemap empty.

Check for Any Console Errors

You can see the list of errors at the browser console, already labeled and color-coded. To address each one of them, make sure you copy the error name and line number correctly. However, you do not always have to manually engage in this process. Knowing the right online audit solutions will save you a great deal of your time:

- **GTMetrix:** helps to check site performance and speed limitations.
  - **Homepage:** https://gtmetrix.com/
- **Search Console:** Google product that is described as: "helps you measure your site's Search traffic and performance, fix issues, and make your site shine in Google Search results."[2]
  - **Homepage:** https://search.google.com/search-console/about
- **SEMrush:** online platform that checks your site in accordance with Google guidelines as well as provides search engine optimization (SEO) and competitor marketing research.
  - **Homepage:** https://www.semrush.com/

Online businesses rest on successfully executed SEO campaigns. Yet without a flawlessly functioning back end that delivers speed and visibility, any marketing tactics would miserably miscarry. Here are some items that ensure SEO-friendly technical performance.

---

[2] https://search.google.com/search-console/welcome?utm_source=about-page, Google

Clean Code

We have already established in the previous chapter that no one likes a slow website. And if you are committed to providing your viewers with a seamless and fast-loading experience, you may start by cleaning up your code that would result in increased site speed. Yet it is not only about speed matter, clean code also makes sure search engine finds and recognizes your site content better.

Clean URL

Another back-end technique that would result in more efficient SEO is optimizing your Uniform Resource Locator (URL), or in other terms, your web address. To be certain that your URL is both engines and readers friendly, have a look at these points:

- Think in advance and identify your target keywords that you are going to incorporate your URL before launching the website

- When spotting keywords for a web address, make sure to avoid keyword stuffing ("practice of shoving as many SEO keywords onto a page as physically possible")[3]

- Use hyphens to divide words (at all times)

- URL has to be readable. Potential viewers as well as search engines must be able to skim over your address name and construe what the content would be about.

It is advised to decide on the right sort of URL before you activate it. Any modification after the site is launched and "out there" has to be convincing and would depend on the Content Management System (URL) that you are working with. And if you decide to change your URL, you would have to create a 301 redirect map to rechannel your visitor from the old web straight into a new one without them having to face the 404-Page Not Found.

Crawlability

Everyone wants their website indexed and ranked properly. For that to happen, search engines release Internet bots called crawlers that help to

---

[3] https://www.wordstream.com/blog/ws/2012/03/21/dangers-of-keyword-stuffing, Wordstream blog

index and store the copy of your site. Ranking and improving the popularity of your web would be impossible without this procedure. You can improve crawlability by regularly managing your site structure (HTTP aspects, metadata) making sure bots have clear access to your content; another option is to check for any broken or bad links that can stay in the way of crawlers doing their job in the right manner and order.

The back end of the website is often seen as less essential compared to the front end. Even though most people mostly interact with the visible layout of the site more, it is important to understand that back and front end work together, in absolute symbiosis. Technical optimization generates a foundation to lay attractive and sound external foreground. And back-end programs have to constantly make an adjustment to ensure the speed, visibility, and indexing of the website. Thinking that technical optimization is too difficult to deal with and avoiding it entirely cannot be a good decision to make. Including back end into the overall SEO strategy can deliver great advantages over other companies. No one should miss that opportunity. You have already started small by reading this chapter, and if it still leaves you intimidated, you can always hire a developer to implement the suggestions mentioned above.

Optimizing Database

Unoptimized databases can also slow affect your back-end performance. To speed up the database, it is recommended to consider using indexes and denormalizing the database. Denormalization here stands for the calculated modification of a normalized database to decrease the time required for select queries by adding redundant data like extra tables or attributes into existing tables to make data more accessible.[4] Database denormalization can help you address the following issues:

- **Big number of joins:** It is often needed to join large numbers of tables in queries to a normalized database. While a table join is a time-consuming operation, such queries use up server resources and take forever to manage. To avoid these issues, you should consider denormalization by adding an extra field to one or few of the tables.

---

[4] https://rubygarage.org/blog/website-speed-optimization-backend, Ruhygarade blog

- **Calculation values:** As a rule, complicated calculations slow down your database's performance. If your database is regularly administering complex calculations as such, it makes sense to add additional columns to a table to hold frequently used and hard to calculate data. Creating a column that contains precalculated values can save time during query execution.

Making Use of Web Hosting Service

Web hosting services are great backup for server management. It not only helps to improve website performance but also offers various capabilities and scalability options. Compared to all web hosting, shared web hosting is the most popular technique. The following three hosting types have enough capabilities like how to improve website performance:

- **Virtual private servers:** both low-cost shared web hosting package and costly dedicated hosting connect the virtual private server (VPS) that extends a personal virtual server and configures shared hosting configurations. The prices of VPS are quite affordable, but if your website requires some additional services it may demand additional costs. It is a convenient solution for an average traffic that occasionally might experience traffic spikes during specific periods.

- **Cloud hosting:** is recommended for small and medium businesses implementing applications with unregulated, unstable traffic like the one of e-commerce websites. With Cloud hosting, you pay only for what you use. Therefore, you need to know the exact amount of resources your website demands and avoid paying any extras.

- **Dedicated hosting:** offers you a dedicated server that is used only by you. Conditions of usage are quite straightforward. First, you are expected to pay the server rent, which is $150 per month. Additionally, you should have a system administrator that can manage your server. With that, you are more likely to get all the power and resources that you need from the operating system to this type of dedicated hosting.

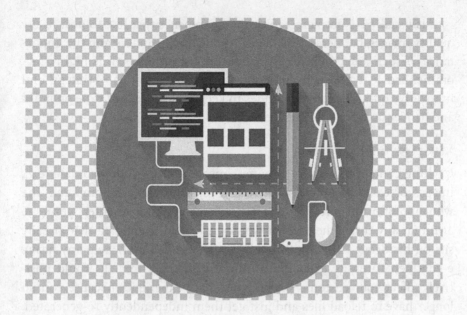

FRONT-END OPTIMIZATION

On-page optimizations could be defined as "adjustments that can be made to the front end of the website, allowing users as well as bots to understand your website. On-page components can be seen by both the users and bots, whereas technical adjustments are just seen by bots."[5] Compared to the back end, there are more elements that could be optimized at the forefront. For instance, files, images, navigation, design, or content. You can adjust and make your on-page as personalized as you wish. The checklist below can give ideas on where you can start:

Keep Site Speed in Mind

Why do you need to keep the website load speed in mind when thinking about front-end optimization? The reason is that placing all the front items like images, fonts, and others take most of the total load time. That is why it is important to focus on smart optimization that would efficiently and timely let viewers interact with your website.

---

[5] https://www.ecreativeworks.com/blog/what-is-the-difference-between-front-end-and-back-end-optimization, Creative

It is usually e-Commerce, multimedia websites that like to post heavy quality images, banners, and videos. And while we understand the importance of such a decision to advertise and sell products, it is highly recommended to back such content with reliable front-end strategy and innovative optimization techniques to keep their page load time on point and deliver user-friendly service.

Make Use of Browser Caching

Caching content in viewers' browsers means storing your web data (e.g., images or cascading style sheet [CSS]) on each viewer's computer. The whole point of this procedure is that when the visitor re-enters your site again (at any time), the content would be loaded using data that already sits on his drive. Utilizing this solution would definitely help any heavy e-commerce website to lock a minimal load time limit. It would positively affect your customers' whole purchase experience because you no longer have to reload files and just get them independently re-generated. Developers identify two types of caching out there:

- **Fragment caching:** refers to boundless caching using fragment caching for dynamic web applications that may quickly become irrelevant.

- **Page caching:** is the most widely used caching for performance optimization, which is wholly implemented on the webserver. After executing on the web server, it returns cached static content to your application. But in some cases, page caching is not suitable for applications having frequent results that would produce content a user has seen before.[6]

- To help you cache your data and therefore reduce web loading time you can use popular software like Squid (caching proxy—http://www. squid-cache.org/) or Varnish Cache (open source HTTP engine proxy— https://www.varnish-software.com/community/varnish-cache/).

Optimize CSS

CSS is a design sheet coding for fonts and styling that makes your web customized and personalized across the whole site. However, having long

---

[6] https://www.bacancytechnology.com/blog/improve-your-website-performance, Balancy technology

and heavy CSS files that need to be reloaded every time someone enters the site can seriously slow down front-end speed. Here are some options you can take in that regard:

- **Merge CSS files:** Merging several CSS into one file can help to reduce the site load rate. By combining CSS, browser requests would come in sync thus reducing the time it usually takes CSS to be delivered one by one.

- **Minimize CSS:** Minimizing CSS basically means reducing the number of codes, spaces eventually making the document shorter and lighter. If you do not know how to start cutting edges, there are many online solutions out there to help you with that. For example, CSS Online Compressor—https://www.giftofspeed.com/css-compressor/, Clean CSS—https://www.cleancss.com/css-minify/ or Online CSS Minifier—https://cssminifier.com/.

- **Adding CSS to HTML document:** If you have small or previously combined CSS files, you can add them directly to the HTML document rather than request it as a separate document each time. This too is recommended as a means of solving issues with slow web loading time.

Optimize Images

Usually, images take up most of the website space, and by compressing, reducing their size with minimal or no visible features loss, you can optimize your front end, making it lighter and faster. If you are hesitant to convert or compress images that take the most space, you can be sure that there are online tools that can modify image sizes without any quality depletion. For example, PageSpeed reduces image sizes to precisely the requested dimension. Homepage—https://developers.google.com/speed/pagespeed/insights/. Optimus.io (https://optimus.io/) does the same job of lossless image compression/decompression. There is a variety of other free options like Optimizilla (https://imagecompressor.com/) or TinyPNG https://tinypng.com/ that can give a compact version of any uploaded image.

Another trick that helps to improve overall perception is the lazy loading of images by using JavaScript (JS) that requests and places images as they become accessible. It is the principle that images get become visible based on the order that they are positioned on the page. So, the main pictures would be loaded first, and the images down below would be the last ones to show up. This trick particularly useful for websites with a great number of images as it creates an impression of quick interaction that is so convincing to viewers; lazy loading also enables easy transfer from one page to another and trouble-free scrolling from top to bottom.

Brotli and gzip Compression

By default, when various items are sent from the server to the browser, they are sent as they are, unchanged. Using compression tools such as gzip and Google's Brotli, we can decrease the file sizes of these requests exponentially:

File	Uncompressed	Gzip	Brotli
style.css	346kb	37kb (-89%)	24kb (-93%)
script.js	106kb	27kb (-75%)	23kb (-78%)
screenshot.png	1.5mb	1.4mb (-7%)	1.4mb (-7%)[7]

You will notice that compared to the CSS and JS files, the compression ratio on the Portable Network Graphics (PNG) is lower. This is because gzip and Brotli are much better suited for text-based files, therefore if you have the majority of CSS or JS files with a high repetition rate, you should consider these methods of compressing.

Manage Third-Party JS

According to Manning Publications, third-party JS applications are: "self-contained components, typically small scripts or widgets, that add functionality to websites. As the name implies, they're offered by independent organizations, with code and asset files served from a remote web address."[8] Websites request third-party JS's for different reasons—to enable personalized recommendations, feedback, or to track their customers' behavior. Yet, at the same time, these JS can significantly slow down front-end loading and provoke an unsatisfactory user experience. Few things you can do to prevent that from happening would be placing JS near the bottom of the HTML document so it does not interfere with other items' loading rate, or you could also minimize or compress JS in the same way you do with CSS file size. Another solution would be to use the defer attribute, like below:

---

[7] https://buttercms.com/blog/front-end-performance-optimization-techniques, ButterCMS
[8] https://www.manning.com/books/third-party-javascript, Manning publications

```
<script src="third-party-script.js"></script>
<script src="third-party-script.js" defer></script>⁹
```

Without the defer you get the render time of 2.2 seconds. With the defer attribute, the load time speeds up to 1.7 seconds, which is 22% faster than before. Apart from the above mentioned mathods, regular inspection of your site to see what JS elements are still being used and what you might need to eliminate can help as well.

Prefetching

Prefetching is a mechanism that allows the browser to store web resources in its cache to then fetch and display that content when the user decides to access it. In other words, "once a web page has finished loading and the idle time has passed, the browser begins downloading other resources. Once a user clicks on a particular link that has already been prefetched, they will see the content instantly."[10] Prefetching can be broken down the following way: a web browser begins parsing and displaying a page; then the browser starts resolving domain names associated with links on the page. And when a viewer clicks a link, the target IP is already accessed and the browser is immediately directed to that server. There are three types of prefetching:

- **Prerendering:** mechanism that actually uploads the entire site the user is expected to navigate to next. Pre-renders web is then stays hidden on the background of the currently open website. Once the viewer decided to move to the net site, it would appear at once fully loaded and displayed.

- **Link prefetching:** permits the browser to fetch presumably expected links of content that the user will request next.

- **DNS prefetching:** Domain Name Server (DNS) looks up for links on a page while the user scans the current page.

Prefetching is sometimes confused with Content Delivery Network (CDN) caching techniques. And while both are involved in site optimization,

---

⁹ https://buttercms.com/blog/front-end-performance-optimization-techniques, Butter CMS
10 https://www.keycdn.com/support/prefetching, Keycdn

they act applying different methods. And it is important to understand that CDN, as opposed to prefetching, operates on the browser side, a CDN stores cached website content using a network of strategically placed points to presence (PoPs).

Optimizing the front end is a continuous process that requires notable analysis and implementation of the compound, innovative practices. And in order to bring a positive impact to front-end development, it has to be frequently tested.

FRONT-END TESTING

Front-end testing is a way for developers and web owners to analyze and examine site performance from a user's point of view. It was not a very popular practice back in the days. But naturally, as web opportunities expanded with the promotion of JS and multiple CSS, it has become necessary to test how technical processes shape front-end perception and what can be done to prevent any errors from happening. Dismissing front-end performance testing might harm businesses in terms of maintaining high operation standards and customer retention.

Unlike front-end testing, back-end testing is not concerned with the application's user interface (UI). It mostly checks whether the database stores the right data entered using the UI. However, the functionality and business processes that make a front-end application work are what make up the back-end server-side layer of a three-tier architecture. Main three benefits that the testing procedure can bring to the table include:

- **Improved user interactions:** Through testing the front-end performance, you would be able to check and, if necessary, reduce the load time, ensure consistency of content and web design across all pages; as well as get the perspective on how you can improve the overall user journey. With all that, you may be certain that the quality and quantity of user interaction will drastically improve.

- **Detecting customer-side performance issues:** The testing procedure will focus on the front end and assess it the way any viewer would—is the interface sequence eloquent and fast? Can you make sense of all the products that the site has to offer? Is the keyword flow correctly timed and located across all pages? Going through the web from a different standpoint might uncover an array of items that need fixing.

- **It is easy and cost-effective:** When it comes to testing toolkits, most applications are available online and free to use. The process itself is easy to operate and does not have to take much of your time.

Types of Front-End Testing

Developers would not be able to deliver stability of a product if they did not engage in testing. Front-end testing covers a variety of practices that have been introduced in back-end development years ago. A testing strategy that you choose to guide the process should be suited to your codebase and testing objectives. To know the best option, you must know the types of front-end testing:

- **Unit testing:** unit refers to the smallest part of calculations and input validations that are testable. Unit testing is the lowest-level testing among the different testing types.

- **Acceptance testing:** with this method, testers test a system for acceptability. They evaluate the compliance of a system with delivery requirements. This type also allows scanning the running application. It helps to ensure the proper functioning of user flows, user inputs, and designated actions.

- **Visual regression testing:** visual regression testing involves capturing UI screenshots and comparing them with previous screenshots. This type of testing is unique to the front end. Testers use image comparison tools to detect differences between the two shots. If you are launching a new website, save this testing process for last.

- **Performance testing:** is of prime importance in the front end that determines the stability, responsiveness, and speed of a product. It also examines how a device fares under certain conditions. Most of the tools available for performance are plug-and-play.

- **End-to-end testing:** is created to ensure that the app behaves according to requirements throughout its running. It also identifies system dependencies, limitations, and enables testers to fix any system issues.

- **Integration testing:** is useful for testing combined units after integration. This testing option exposes all kinds of errors that may occur after merging codes.

- **Cross-browser compatibility testing:** focuses on allowing users to have the same experience on different browsers. It is helpful to ensure that an application works correctly on different devices and browser combinations.

How to Create a Front-End Testing Plan

A front-end testing plan should reflect the project needs and objectives. You can start working on your own plan following these guidelines:

- **Decide the budget:** it is important to decide the budget before running test demos. You should relocate your budget according to the timeline of tasks. And even if you find yourself on a tight budget, the quality of your testing tools must be prioritized.

- **Decide the tools:** once you have set the budget, you can make a list of all the tools you will be needing. There are many tasks involved in front-end testing. So you can also use specific tools for different tasks or you can opt for a single tool that performs multiple functions.

- **Set a timeline:** any project needs a clear time framework. Especially with front-end testing you might find yourself covering many aspects. Therefore, to avoid a major time constrain you need to set a timeline before you start testing.

Front-End Performance Testing Tools

Now that we have established a strong case for conducting front-end performance testing, it is time to look through what testing tools are there to use. Ideally, the choice of the testing solution has to depend on your project objective and the data that you want to gather about the site in the end. A basic testing application as offers real-time monitoring, statistics about the performance of each item of the page, and insights in regards to overall performance. Here is the list of the most widely used testing tools at the moment:

1. **Lighthouse:** Integrated into Google Chrome, the application assesses the accessibility of web pages (60–90 seconds per page) and performance-related insights. Most importantly, at the end of testing, it

files a complete report presenting findings and audit summary, listing all detected issues and sharing potential solutions.

Homepage—https://developers.google.com/web/tools/lighthouse/

2. **Web Page Test:** It is one of the most feature-rich platforms that offer a range of latency filters and comprehensive network visualization. The tool can be used across different devices—Mac computers, tablets, iPhones, and Android smartphones and has an authentic interface display.

Homepage—https://www.webpagetest.org/

3. **Pingdom:** Pingdom is another real-time front-end performance monitoring that is widely used in companies like Netflix, Amazon, and Spotify. Tech communities choose it for its features like continuous page load time tracking, parallel interaction monitoring, and rich performance data.

Homepage—https://www.pingdom.com/

4. **Httperf:** Httperf is a solution for web-server testing that provides a full set of metrics for performance management. The tool aims to design micro-and macro-scale benchmarks in integration with other workload generators and sustainable server overloads.

Homepage—https://github.com/httperf/httperf

5. **JMeter:** JMeter is an online tool for automated front-end testing that supports a wide range of protocols and languages. JMeter operates as a simulation—by imitating to be a user, it sends a target request to the server and reports back with data analysis about the website's overall functionality. And even though it might sound complicated, it has a very easy-to-use interface and open-source license.

Homepage—https://jmeter.apache.org/

Regardless of your project's site size and scale, it is still recommended to carry out front-end testing on a regular basis. It would be particularly beneficial for the following groups:

- **Companies with the high-traffic website:** The high-traffic website generally experiences crashes during times when traffic spikes beyond the limit. With testing, business owners will be able to prevent unpleasant occurrences as such and proactively manage traffic hours.

- **Small business owners that want to get more search ranking exposure:** Small businesses with modest marketing budgets will be able to benefit from front-end optimization as it would evidently improve the most important interaction metrics like speed and site frame. By advancing just these functions, business owners will attract more audiences and upgrade their web's rank.

- **E-commerce store managers:** Front-end testing can draw attention to malfunctioning e-commerce site items like browsing the catalog, buyer's cart, digital check-out counter, and transfer of funds. Once optimized, satisfactory speed and improved responsivity level will directly affect the store's conversion rates.

CONTENT OPTIMIZATION

People produce content of all types at an extremely accelerated rate. You are providing search engines with data that later would be generated in keywords search results through content that you post. Knowing that, you want to tailor your content to attract the audience you are aiming for. And if your content is not compelling enough to serve the overall marketing strategy, you need to optimize it. Developers identify four key web content formats—text, image, video, and news. We are going through each one of them in this chapter.

Optimizing Text

Text content you choose affects website conversion rates even before viewers land on your page; it starts from the moment they read your URL and see the title tag you used to describe what should be expected as the main content.

- **URL:** A catchphrase URL provides search engines with more information that enables them to place you correctly in search results. When optimizing your URL structure, look for relevant to your content keywords and phrases that would give a direct and suggestive perception about the website. Shall you have more specific content you may want to include straightforward and long-tail keywords that will make your site stand out?

- **Title tags:** Or page's title tag gives the audience a brief description of what could be the most important to know about the page. Page tag is considered to be one of the principal elements of website optimization for a reason—title tags appear in numerous places to outline and give details of the web content. For instance, search engines use them to link titles to search results display; same page tags also appear on top of the browser when the page is viewed. In addition, when you share the website content with someone through social media, the title tag is exposed there as well. Consequently, optimization of some sort might be needed just to make sure page tags serve site promotion in a significant manner.

- **Meta descriptions:** Meta description is a key piece of information that search engine pages usually use as the short specification for each result; it is also the first thing that web users see when they click to view your page. Successful meta descriptions have to accurately state the intent of the page content and include related and most appropriate keywords. "Overstuffing" and pasting as many keyword tags as possible will only hurt your ranking. It is best to prompt meta keywords by selecting only truly traceable terms.

- Keep in mind that it has to be concise as search engines likely to cut off any presentation that exceeds 150 characters. It is also advised to avoid general and unbranded descriptors that do not introduce anything special about your website.

Optimizing Images

Viewers are interested in images almost as much as in text. Thus, the visual content you choose to display on your website has to be regularly reviewed for more clicks. Depending on the kind of images you want to see, you can apply different formats. Not to put too fine a point on it, we recommend to:

- use JPEG for larger illustrations and images

- select PNG to keep background transparency in the image

- use Scalable Vector Graphics (SVG) for logos and icons

Since search engines can't acquire full information from image-based content just yet, they rely on alt tag and title tag as well as file names to derive contextual meaning and whether it is of any relevance to the search. Therefore, it is essential to optimize image tags to produce better, organic results to get more views.

- **Alt tags:** Alt tags are used to describe the image when the image is not accessible to be displayed. This usually happens when the viewer's Internet connection is slow and the web cannot load images correctly. Wikipedia explains the importance of having catchy yet simple alternative descriptors in the following way: "In situations where the image is not available to the reader, perhaps because they have turned off images in their web browser or are using a screen-reader due to a visual impairment, the alternative text ensures that no information or functionality is lost."[11]

- **Image tags:** Same as alt tags, image tags help users comprehend the image's context by locating specific words and phrase combinations that appear when scrolls over an image or gallery.

---

[11] https://en.wikipedia.org/wiki/Wikipedia:Manual_of_Style/Accessibility/Alternative_text_for_images, last edited October 19, 2021

Alt and title tags are impacted within the image code in this way:

```
<title="tag types in content optimization"
alt="tag types" img src="tagtypes.jpg" />.
```

- **Filename:** Fitting filename is important to provide requisite information that would make the relation of the image to the other content on the page clear. Similar to images' alt and title tags, proper image filenames should consist of the keyword that you're optimizing your content for. Thus, instead of uploading your image with a filename like "DC000111.jpg," insert some descriptive titles, like "filenamematters.jpg."

- **File size:** According to HTTP Archive of November 2018, images take around 21% of a total webpage's weight.[12] So when thinking about optimizing a website, image size might be the place to start. No one would be able to see alt tags or image tags if the page will not simply load. Yet visual optimization can be just as easy to edit as tuning fonts or file names. As mentioned previously, numerous online solutions can help to adjust the image size for free without having to sacrifice the visual quality.

Optimizing Videos

With recently developed social media applications, it has become trendier than ever to produce video content. From the many videos that are popular on the Internet, we can learn that posted videos are not at most times professionally recorded or masterly scripted, but just funny and entertaining. Launching a branded channel on the video site of your choice could bring more recognition and give your business a certain modern edge. And once you create high-quality, amusing, and well-optimized video content, big names in the video broadcasting industry like Youtube or Vimeo might want to host it as well. For that to happen, better pay attention to the following items:

- **Video titles:** With video titles, you want to make sure that the name you create would be catchy enough to make viewers want to watch it till the end. It should contain relevant keywords, but also intrigue

---

[12] https://httparchive.org/reports/page-weight, HTTP archive

and sound inviting to your ideal audience. Unexciting and flat titles that are overly calculated or "overstuffed" will lead visitors to believe that your video is just a part of a manipulative marketing campaign.

- **Video descriptions:** Good description should prep and clearly picture what the video is about, setting the right expectations from watching. Principally, it should be constructed to address and be directed to your audience first, and search engines second. It would also be useful to add a link to your main website in the video descriptions so viewers can access additional content and contact points you provide.

- **On-site optimization:** Just like with images, posting suited and optimized title tags will give more context to search engines to fully understand, rank, and suggest your content.

Optimizing News

Any website owner nowadays could be viewed as an independent publisher as they have the platform and ability to produce any sort of thematic report or review. Even if it is something as personal as blogging or as big as running live news, optimizing content can bring wider recognition through the availability of social media outlets. However, online news is considered to be a very competitive space that demands generating only engaging and intriguing content. This implies that in order to attract readers and rank high on search engines, you need to broadcast and optimize everything that originates on your web. Basic considerations involved in news content optimization involve the following:

- **Headlines:** To put it simply, in order for your news to rank high in search results, the article headlines should be stimulating and tempting at the same time. To achieve this, you need to know what keywords, phrases, and events attract most readers' attention recently. Staying on track and being aware of worldwide trends and occurrences will help to capture the traffic rate you potentially aiming for. And to publish news content that can compete with other news portals, you will need to create each article on its own separate page that could exist on a static URL.

- **Keywords:** Ideally, news articles should attach a well keyword-optimized description. Primary keyword material should be coming from your article only, yet avoid "overstuffing" the web content thinking it would bring the right traffic. In addition to text, if your news piece includes photos or video elements, make sure to optimize each according to the guidelines indicated earlier.

- **Creating quality:** The process of ensuring quality content demands consistency. It has many big and small elements that nevertheless equally important:

  - Proper grammar

  - Correct spelling (typos can ruin the flow and deeper meaning of your article)

  - Regulate formatting (add bullet points, highlights or break your content into distinguished paragraphs)

  - Adjust the length (no one has enough patience for long-reads these days; you might want to cover the topic thoroughly but make sure to adjust the content length to your reader)

  - Be readable (do not wander off your content focus and keep the same tone throughout the article)

  - Be clear (readers should not guess your point, deliver in an engaging yet reasonably obvious manner)

  - Be objective (objective pieces help to build trust with the audience; avoid sounding overly assertive or forceful, aim to keep balanced content style)
    To sum up, it is indisputable that quality content matters. If you fail to originate well-adjusted, interesting, and authentic content, you might let your readers down and lose their trust.

- **Establishing content authority:** It turns out to be hard to define what an authoritative article is, yet you can almost instantly identify one when you read one. Drafted below are several features that can help you to create optimized and authoritative content:

  - Make sure you possess enough expertise to write about things that you are covering. Have you researched related data before posting it? If you're writing about coding, your readers would presume you have done it your whole life. And if you're writing about modeling, then you must be Naomi Campbell, no less. This is just to illustrate that in case your viewers find a flaw or minor inconsonance that suggests your lack of knowledge on the matter, your content will lose its authority.

- Citing authoritative sources will help your content to establish its own authority. You almost never create an authoritative article using only your opinions. Referencing other sources is considered to be a standard practice. The library guide at the University of California Berkeley formulates it like this: "Citations provide evidence for your arguments and add credibility to your work by demonstrating that you have sought out and considered a variety of resources."[13] Adding details on respected research centers, consumer survey results, or insights from economic data or interviews with related authorities can also improve your content quality.

- Inviting authoritative authors, experts in particular fields, or leaders of opinion to write for your website can enhance your content credibility as well attract new traffic that might come with them.
Features like likes, comments, and social shares can be used as an indicator of your content gaining certain certitude. But establishing content authority should be perceived as a marathon, not a sprint. And just like most in life, it takes time. You cannot expect to gain trust after just a couple of decent articles, it takes more invested commitment than that.

- **Producing calls to action:** It is known that well-optimized content is typically action-oriented. A call to action means a content intent that prompts the reader to take preferred steps and bring conversion. If you create a piece of content, you most likely want to insert a specific call to action that lets the user know what to do next. This is true for YouTube videos, any social media posts, blog posts or podcasts, and many others. With a correctly formulated and placed

---

[13] https://www.lib.berkeley.edu/research-support/cite-sources, UC Berkeley

call to action, you can convince people who visit your website to do something specific like read the rest of your article, share content on social media, sign up for more updates, download site content of buying a product.

There are few tricks to make calls to action more inviting:

- Directly insert calls to action in the text of your content

- Install pop-up advertisements to encourage conversions

- Position call to action in the web's sidebar

- Use popular website boosters like HelloBar to prompt conversions

And if you are wondering whether your content optimizing practices have worked successfully, you need to check whether the number of your conversions has increased since the implementation.

ADDITIONAL OPTIMIZATION TIPS

Optimizing your website is an important project that would require every bit of your creativity and attention to detail. The good news is that the potential reward is considerable and worthwhile all the effort. Prepare to test, fail and start from scratch if what you are aiming for is perfection. Below is the summary of the top six technical errors or inaccuracies that you want to avoid to save your time:

Always Backup Your Site

The web optimization mantra for every developer or site owner to repeat every day and hour is: always complete a backup before implementing any changes. After you have found the problem but before you ready to make changes, always backup your website. If you use a third-party testing tool to A/B test changes, you can do it after the testing phase. But in other instances, you need to backup before making the changes live.

If your site is using WordPress, you can utilize a plugin like UpdraftPlus (https://updraftplus.com/) to back it up and avoid any errors. If you run your own Content Management System you could download third-party solutions like Drop My Site (https://www.dropmysite.com/) or cloud backup software like Carbonite (https://www.carbonite.com/).

Optimize Your Images before Uploading

As mentioned before, images can easily be the heaviest part of a web page and directly impact your loading speed. To make sure your users do not face frozen or snail-like web you need to optimize all your visionary content before uploading. There are many free, online solutions that we have touched upon earlier in this chapter. It is also recommended to merge together background images into one connected picture as it will minimize the number of separate HTTP requests your site makes.

Use a CDN to Improve Page Speed

You can also improve the image load speed by hosting all your media files as well as scripts through a CDN. The definition and main advantages of choosing CDN described by Globaldots Network Provider:[14] "is a globally distributed network of web servers or PoP whose purpose is to provide faster content delivery. The content is replicated and stored throughout the CDN so the user can access the data that is stored at a location that is geographically closest to the user. This is different (and more efficient) than the traditional method of storing content on just one central server. A client accesses a copy of the data near to the client, as opposed to all clients accessing the same central server, in order to avoid bottlenecks near that server." Thus, using the closest servers to load files makes data travel time short and significantly speeds up page loading time. Two of the most popular options of CDN at the moment that you should take a look at are Cloudflare (https://www.cloudflare.com/) and Amazon CloudFront (https://aws.amazon.com/ru/cloudfront/?nc=sn&loc=0).

Test Often

Instead of having to test all items in one mega-test at once, it is strongly advised to set a schedule for the regular single-element test. That way, you not only will be able to identify individual errors but also have a short and long-term optimization plan to follow.

Optimize for People

When selecting SEO and content analysis tools, you need to think about how you would balance between delivering for Google search engines and

---

[14] https://www.globaldots.com/content-delivery-network-explained, GlobalDots

offering good value for your viewers. If you make the basic mistake of focusing only on one side of the equation, both your ranking and conversion rates will be troubled.

Keep Mobile in Mind

Mobile-first is the dominating developer strategy in the current era. With heavy Internet traffic coming by and large from smartphones, it is crucial to test your website for operating in different devices and optimize keeping the mobile in mind, focusing on mobile-friendly customer service, smooth purchase, and post-purchase transactions.

CONCLUSION

Website optimization is an evolving and continuous process of constant advancement by learning new information about the behaviors of your visitors and developing working theories that distinguish and constitute different parts of website optimization. This particular domain is very broad and may involve acquiring a number of professional skills from development to design to be able to reform every item in web content.

In the past, it might have been enough just to have a website for your business. But with the current globalization and high competition levels, you are pressured to regularly optimize your site considering a variety of factors like SEO and customer interaction that can affect not only your ranking but the future of your business altogether.

We have reviewed two primary operations that responsible for web maintenance and optimization—the back end and front end. Both need to work hand in hand as there would be no traffic if your back engines are defective and malfunctioning; yet even technically mega-advanced site will not attract conversion rate if the displayed content is outdated and simply dull. Content optimization should be treated as an integral part of web development as it is accountable for optimizing elements that bring style like image, video, text, and news content.

Having this information should make web testing and optimization relatively easy for you. Once you succeed in every area of web practice, you will certainly see the transformative impact that it will cause. You can start by creating your optimization plan using our guidelines and tips to avoid common mistakes and get insights on how you are can lead this process.

CHAPTER 3

Web Performance and Speed

IN THIS CHAPTER

➢ Web Speed

➢ Speed Metrics

➢ Optimizing JavaScript

➢ Minification

In the previous chapter, we discussed several web optimization techniques and methodologies. We divided such optimization methods on the basis of front end and back end.

Beyond that, we also covered several optimization ideas and strategies related to content in itself. As far as the content part is concerned, we will be revisiting localization, search engine optimization (SEO), and other content-related optimization ways later on during the course of this book.

Similarly, for back-end optimization, especially server-side compression and likewise, we will come back to it in subsequent chapters.

In this chapter, we will be focusing our attention on web performance and speed in itself. As such, we will be covering certain technologies and ways in which we can boost the overall performance of our web applications and websites. Beyond that, we will also be looking at speed optimization in good detail as we progress through this chapter.

DOI: 10.1201/9781003203735-3

51

Considering the fact that we have already discussed the importance of web performance and speed from the perspective of web optimization back in Chapter 1, we can safely bypass that discussion and jump straight to performance and speed improvement ideas and tools.

It is worth noting though, it might be a good idea to read this chapter in assonance with certain parts of Chapter 6, as several web performance and speed-related topics are also mentioned therein, especially when it comes to optimizing media elements, code snippets, and images for faster loading.

That said, let us first get started with the most obvious entity in this section—Content Delivery Networks (CDNs).

CONTENT DELIVERY NETWORKS

Everyone talks a good deal about CDNs nowadays, and the options are plentiful. But what are these CDNs, and how can we make the most out of them as far as web performance optimization is concerned?

What Is a Content Delivery Network?

A CDN is a conglomerate of geographically distributed servers and data centers for spatial distribution of data to users around the world.

Here is how Wikipedia describes a CDN:[1]

> A content delivery network, or content distribution network (CDN), is a geographically distributed network of proxy servers and their data centers. The goal is to provide high availability and performance by distributing the service spatially relative to end users.

In simple words, a CDN works by putting together a range of servers and data centers around the world. Now, as and when a given user accesses a web page, the CDN fetches it from the delivery network or data center closer in proximity to the user's geographical location.

Naturally, this cuts down the latency and the time required in transmission of data between geographically distant locations. As such, CDNs can help boost web speed by keeping a cluster of servers and data centers and using geographical location to their advantage.

---

[1] See: https://en.wikipedia.org/wiki/Content_delivery_network, last edited on November 26, 2021

This, of course, is not the only advantage that a CDN brings to the table. CDNs can and have proven especially useful when it comes to website security and hardening, as a CDN can keep spam bots and DDoS attacks at bay, if properly configured.

We will be coming to the advantages that CDNs have to offer in a short while. Before that, let us first understand how a CDN works.

How Do Content Delivery Networks Work?

We know that a CDN, in itself, is a wide network of various servers and hardware platforms, each is marked by a distinct geographical location. Now, considering the fact that nearly every website on the Internet is made up of HTML, CSS, JavaScript, and other multimedia assets, a CDN leverages this fact and creates handy copies of such assets.

Thereafter, the users are served cached copies of the assets by the CDN, depending on their location in the world and the server that is nearest to them in terms of latency issues and proximal distance.

To facilitate this, CDNs connect to various Internet Exchange Points (IXP).

"An IXP is a key location where various Internet service providers and networks interconnect to exchange information and to request and/or offer access to each other's networks for global coverage."

By ensuring that it is directly connected to the IXPs, a good CDN can cut down transit time of data and help in faster web performance.

The next aspect in the functioning of a CDN is that it utilizes a range of strategically placed servers and data centers across the globe. Depending on the requests originating from a given location, a CDN provider might deploy less or more number of servers therein. For example, there are likely to be more servers and data centers in the Asia Pacific region as compared to the Sub-Saharan region, simply because of the number of requests and data transfer calls that originate from the former is greater than those originating from the latter.

To sum it up:

- CDNs tend to establish direct uplinks with various IXPs for faster data transfer.

- CDNs distribute their servers geographically, and when a user attempts to access a given web page, the CDN fetches and provides its own copy of the page, from a closer data center, instead of pulling the page from the website's origin server that might be far off geographically.

A good CDN tends to have a large network of servers across the globe. Beyond that, many CDNs also offer various other web optimization tools and tactics, such as file minification, code compression, and so on.

Even more importantly, modern-day CDNs come with additional security features, such as Web Application Firewall of their own, protection against DDoS attacks, as well as ability to detect malicious spam bots and traffic and block them therein. In this manner, CDNs can help in saving crucial website bandwidth, especially if our website is hosted on a limited bandwidth plan.

Now that we have already started discussing the benefits and features that CDNs have to offer, it might be a good idea to take a look at the various advantages that CDNs can provide us with.

Benefits of Using a CDN

If you aren't already doing so, you should by all means consider using a CDN for your websites right away. The advantages or benefits associated with a CDN are numerous.

First up, the most obvious advantage that a CDN has to offer is the speed. It can greatly help us improve the speed and web performance of our websites by fetching the assets from closer servers and rendering web pages much faster.

Secondly, CDNs are famous for helping in reducing the load on the actual origin server where our websites might be hosted. Since several requests and assets are being served from different servers, a good deal of bandwidth and memory resources are saved on the origin server. This makes life especially easier for folks who might be using a shared hosting package, or a less powerful hosting solution.

Thirdly, the majority of the CDNs nowadays tend to come bundled with security features. These can be as simple as presenting Captcha challenges to suspicious visitors, or as diverse as handling Web Application Firewall and offering DDoS protection. By filtering out the good users from the bad ones, not only can a decent CDN help us ensure our website remains in good health and is not compromised by malicious users, but also add a pinch of accuracy to website stats and metrics and fake bots are blocked by the CDN.

Certain CDNs, such as CloudFlare, also offer a reliable level of data encryption. This can act as an additional layer of security for our websites.

Fourth, some of the higher-end CDNs have redundant architecture. This means if a given web server is having issues, users and visitors can automatically be redirected to an alternate server. In fact, even if the origin server is down, a cached copy of the web pages can be served by the CDN at times. Of course, certain functionality such as signups or AJAX scripts may not be available in cached copies of data, but it is still better than "website down" status that might otherwise be shown to visitors in the absence of a CDN.

For websites and web projects that run into niche areas such as video hosting or streaming or something similar, a CDN is not a luxury but a necessity. This is primarily because streaming hosting a video or a podcast from just one server might make it nearly impossible for users that might be located on the other side of the world. A CDN can ensure that everyone can access video and audio content with equally reliable and good speed, thereby affecting the overall web performance and speed positively.

As can be seen, CDNs can go a long way in improving web performance and overall speed of our website.

How to Pick the Perfect Content Delivery Network?

Finding the right CDN is as important as, say, finding the right web hosting provider or domain name registrar. Considering the fact that a good deal of our website's performance depends on how good our CDN actually is, it becomes all the more important to find the right CDN that is suited for our needs.

Essentially, the decision as to which CDN is better than the others depends on three major factors:

1. The nature of our web project

2. The budget in question

3. The features that we actually need

Nature of the Project

When we talk about the nature of the website or web project, we are not necessarily referring to the niche or genre. Although, the niche also has a role to play in this—as we discussed earlier in this section, video and podcast sites tend to almost compulsorily require a CDN. On the other hand, a simple blog may just do without one.

However, speaking of the nature of the project that we have in hand, we need to first figure out who our target audience is and where might they be located. If, for example, we are running a website with a global audience, it might be a good idea to opt for a CDN that has more of a global coverage, such as MaxCDN or CloudFlare. On the other hand, highly localized audiences, such as those pertaining to a specific region or city, might require us to select a CDN that has better architectural and hardware presence in that particular region.

Similarly, the nature of our content too plays a vital role. If we are running a website pertaining to, say, Baseball, opting for a CDN that has a better presence in the North American region can be a better idea, since that's where most of the Baseball fans reside. That said, won't a Baseball news site probably require frequent updates of scoreboards and other stats? This implies our CDN should be perfectly adept at handling JavaScript and AJAX calls—preferably even retain its own copies of said assets, and also be able to smartly distinguish which assets ought to be static, and which need to be fetched more often.

Budget in Hand

This one is fairly simple. Can we afford to invest in a CDN for a regular period of time? If not, it might be a good idea to rely on a free CDN in lieu of a paid one.

Generally, most of the A-tier CDNs, such as the above-mentioned CloudFlare and MaxCDN, come at a premium price that is billed per month, depending on how heavy our usage might be. Certain CDNs also offer a free plan with limited features that can serve the needs of a small to medium-sized website or web project fairly well.

Nonetheless, a CDN can go a long way in improving the overall web performance and speed of our website. As such, investing some funds in a CDN is a wise idea.

With that said, what if we are on a shoestring budget? There isn't much cause to worry. Unlike the web hosting world where "free hosting" essentially implies poor quality performance and a higher amount of downtime, there are several premium-quality CDNs that we can avail for free. In the next section, we will be discussing some worthy CDNs and therein we will cover some of the best ones that do not cost a dime!

Required Features

At the end of it all, our decision to choose (or not to choose) a given CDN boils down to the simple query—what features do we really need?

This one can, in fact, be further broken down into a series of smaller sub-questions:

- Do we really need a globally diverse cluster of data centers, or is our audience more concentrated in certain areas and we do not need global scalability?

- Do we need the added security features that come with a CDN, such as DDoS protection, and likewise?

- What about the actual usage? How much bandwidth might be required?

- More importantly, what does the website really need? Do we need to stream and store videos? Or are we seeking a CDN that can help us cache and serve our website assets, say code files and other stuff, faster to users?

No two fingers are identical, and neither are two websites. As such, we need to carefully analyze our own requirements and what our web project really is in need of. For instance, a political website cannot survive for long without certain web security enhancement, and if the CDN can boast of some extra security measures, it's a boon for us.

Similarly, different websites might consume different amounts of traffic and bandwidth, and also require different levels of caching and tweaking. Once we have figured out the exact needs and requirements of our website or web project, we can then head on to finalize and select the perfectly suited CDN for our needs.

Popular Content Delivery Networks

At this stage, we have learned what a CDN is, what it does, and how it can help us improve the performance of our web projects. Beyond that, we have also covered certain key aspects that can help us adjudge the better-suited CDN for our needs.

Now, the big question—which CDNs are really worth considering? In this section, we will be discussing some of the major and most popular

CDNs out there, as well as their Unique Selling Points (USPs) and important features that they have to offer.

CloudFlare

CloudFlare is, quite possibly, among the most popular names in the world of CDNs when it comes to scalability of use. Owing to its free and paid plans, CloudFlare has risen in popularity and is especially a big hit among smaller to medium-sized websites and webmasters.

CloudFlare is fairly well known and is nowadays a staple feature in the offerings of most modern-day shared and reseller hosting providers. CloudFlare boasts of a wide range of features, such as:

- Worldwide network of servers

- Integration with innumerable tools, such as WordPress

- Protection against malicious attacks and spam bots

- DDoS protection

- Details metrics and analytics

CloudFlare offers a suite of plugins for various service providers and content management systems. For what it's worth, CloudFlare does make for a very decent and smart choice when it comes to picking a good CDN for your web projects.

With that said, of late, CloudFlare has been eyeing a shift toward the enterprise clientele, such as remote teams and likewise. Perchance you are an agency or a developer who works mostly in a B2C environment, CloudFlare may appear to be too bloated in terms of features (or perhaps too confusing) for you.

Homepage—www.cloudflare.com

Google Cloud CDN

Google Cloud CDN is another popular option when it comes to CDNs. It is fairly obvious that Google Cloud CDN offers seamless integration with the Google Cloud platform. As such, if you are a Google Cloud user, using the accompanying CDN is a wise choice and can save a lot of time, efforts, and energy.

That said, Google Cloud CDN is more of an apt fit perchance you are hosting or streaming videos or handling big amounts of data transfer. The pricing is fairly generous and depends on the amount of data in question, but it varies depending on the geographical region. For instance, the price per GB for the North American region is up to $0.08 for the first 10 TB, whereas it is $0.20 for the first 10 TB for China.

Naturally, Google Cloud CDN may not suffice for everyone's needs but for developers already working atop the Google Cloud platform, this particular CDN is more or less the default choice.

Homepage—https://cloud.google.com/cdn

StackPath

StackPatch CDN is geared specifically toward web speed optimization. Back in the day, there used to be a worthy CloudFlare competitor, MaxCDN. Ever since MaxCDN merged with StackPath, the latter has risen in stature in its segment.

StackPath does have a viable architecture with global presence, and can prove useful when it comes to web speed optimization.

That said, it also has certain interesting array of features, such as:

- Highly customizable set of redirection rules, including custom 301 redirects, etc.

- Ability to add or remove custom headers to requests and responses

- Stellar cache management tools for web cache delivery and retention

- Free SSL certificates

- Support for WebSockets, etc.

In practical terms, StackPath is a tailor-made solution for websites and web applications. The pricing, though, is feasible only if one were to sign up for the various StackPath offerings, such as Web Application Firewall, DNS, and so on alongside CDN. Solo CDN plans do exist but are not the most budget-effective at the moment.

Homepage—https://www.stackpath.com/products/cdn/

KeyCDN

KeyCDN is a CDN that just works seamlessly out of the box and does not seem to have a lot of confusing options attached to it. In fact, unlike many other names on this list, KeyCDN is more of a niche product in the sense the CDN itself is its main offering, not just one among many other offerings or solutions.

KeyCDN also brings to the table several standardized features, such as:

- DDoS protection
- Multiple clusters of cloud servers spread globally
- TLS certificates
- Advanced security and cache features

Among other things, KeyCDN also has its own image compression and caching mechanism, that works by means of URL parameters and can serve images on the fly. For websites such as news portals, photoblog, and other image-focused content, KeyCDN is quite possibly the de facto standard pick.

Homepage—https://www.keycdn.com/

Amazon CloudFront

Amazon CloudFront is a CDN that can be used to deliver data globally. The USP of CloudFront is that it is neatly integrated with other Amazon Web Services (AWS) offerings, and as such, if you an existing AWS user, CloudFront is not only the easiest choice, but possibly also the cheapest.

This is mainly because data originating from AWS offerings, such as Amazon S3 or likewise, is not counted toward the paid quota. As such, CloudFront can prove highly cost-effective for users of AWS.

It is worth noting that AWS has a free tier as well. Users on the free plan can consume 50 GB data transfer out, 2,000,000 HTTP and HTTPS Requests with Amazon CloudFront. In practical terms, for smaller projects, CloudFront is like a blessing.

Homepage—https://aws.amazon.com/cloudfront/

CacheFly

CacheFly has been around for quite a while by now—since 1999.

At a general look, it seems to offer nearly the same set of solutions as most other CDN providers out there. This includes stuff such as security offerings, DDoS protection, image and web cache, a global cluster of servers, and so on. However, there are certain things that set CacheFly apart.

To begin with, it has specialized plans for low latency video streaming or podcast hosting. Other than that, perhaps the most noteworthy of CacheFly's offerings is its "China Acceleration" solution.

For folks looking to get into the Chinese market (or those who are already active there), CacheFly has a China-specific plan that comes with features such as:

- Local ICP license (ability to register. cn domain name)

- In-country DNS (perchance the Chinese government blocks foreign DNS and renders your site inaccessible)

CacheFly claims to be the fastest CDN in China. Claims apart, the above offering of ICP license and other stuff can prove handy for entrepreneurs and companies looking to tap the Chinese market.

Homepage—https://www.cachefly.com/

Imperva Incapsula

Imperva Incapsula is more of an application delivery platform, and less of a CDN. It caters almost entirely to an enterprise-level client base. That said, it does have a CDN offering, that comes with all the standard features that can be found within a CDN, including a global network of hardware and servers.

Nonetheless, it is well worth stressing that Imperva in itself is an application delivery system. As such, it is more geared toward web apps and similar solutions—such as Fintech applications or wallets.

As a result, their CDN too is more apt for such applications and tools, and not entirely the wisest pick for general-purpose websites or similar projects.

Homepage—https://www.imperva.com/products/cdn-content-delivery-network/

Jetpack by WordPress

Jetpack is a WordPress plugin by Automattic, the parent company behind WordPress. The basic idea is to provide the same set of features as WordPress.com to self-hosted WordPress users as well.

As such, Jetpack comes loaded with various features, such as social sharing buttons, photo galleries, contact forms, and so on Among other things, there is also a CDN that allows your self-hosted WordPress site content to be served via the WP.com global CDN.

Here is how Jetpack describes its CDN offering:[2]

> Content Delivery Network for WordPress Sites—Accelerate your site with faster images and static files. Reduced load times for your readers, less bandwidth from your host.

Jetpack CDN is a free offering and works perfectly well with photos and static assets such as CSS and JavaScript files. That said, it is worth noting that it does not boast of various other offerings that certain other CDNs on this list might offer.

Furthermore, it goes without saying that Jetpack CDN, much like all other tools in the Jetpack armor, works only with self-hosted WordPress sites. Users of non-WP platforms cannot leverage the benefit of Jetpack CDN for their websites or projects.

Homepage—https://jetpack.com/features/design/content-delivery-network/

Sucuri

Sucuri is more of a web security company and less of a CDN provider. It has a stellar list of security offerings, including website cleanup and a popular and reliable Web Application Firewall.

The Sucuri CDN is not the world's most feature-rich option to talk about, but for what it can do, there are caching options and various other handy offerings. In fact, the CDN by Sucuri is best used in association with their hardening and web security packages. In other words, Sucuri CDN is not the company's primary offering, and should not be viewed as such.

However, if you are in need of a security or hardening solution, Sucuri is well worth a look. In that case, the added advantage of a CDN might just fit the bill perfectly.

Homepage—https://sucuri.net/website-performance/

Rackspace

Rackspace, just like Sucuri, too has its CDN, but it is not the flagship product. Rackspace is primarily a cloud hosting company, and the CDN in itself has a praiseworthy network but appears to be more of an afterthought.

---

[2] See: https://jetpack.com/features/design/content-delivery-network/ Web, Jetpack

There seems to be no DDoS protection, nor are there any other major features to talk about. For what it's worth, Rackspace CDN might only be worth consideration perchance you are an existing Rackspace customer and content delivery is your prime focus, not the other bells and whistles that most other CDNs have to offer.

Homepage—https://www.rackspace.com/openstack/public/cdn-content-delivery-network

Fastly

Fastly is a slightly lesser popular, albeit equally feature-rich CDN that offers competitive pricing as well as a good coverage of hardware infrastructure worldwide.

Here is what Fastly CDN brings to the table:

- Instant purge and cache features

- Global cluster of servers

- Improved page load times

- Solutions for image optimization and video streaming

Image—fastly.png [Fastly is a lesser known but equally impressive CDN with a global cluster of servers]

Fastly, as the name suggests, focuses heavily on faster page load times. That said, the security features or stuff like DDoS protection are not its strongest forte. However, when it comes to page speed optimization, Fastly is a viable alternative to the likes of CloudFlare or StackPath.

Homepage—https://www.fastly.com/

MetaCDN

In the world of CDNs, MetaCDN is a comparatively smaller and much less famous name. It offers CDN solutions to businesses and freelancers alike.

It is worth noting that MetaCDN's primary offering seems to be its solution for live streaming of video. It intends to provide a reliable and robust solution for anyone looking to stream videos and playlists online.

A good feature is the rollover of monthly credits, such as that any leftover bandwidth is incremented to the next month. For folks on a budget, or for people who are unsure about their bandwidth and data consumption, MetaCDN can prove highly cost-effective.

But at the end of the day, MetaCDN does not seem likely to be anyone's first choice when it comes to choosing a CDN. Of course, this is a subject remark, but it is well worth admittance that owing to the free tiers offered by the likes of Amazon and CloudFlare among others, smaller CDNs may not appear enticing enough for everyone.

In any case, as a niche CDN, MetaCDN caters well to the video streamers, and if your project falls in that category, MetaCDN is worth taking a look.

Homepage—http://www.metacdn.com/

Akamai
Here is how Akamai defines itself:[3]

> Akamai is the creator and operator of the world's most highly distributed CDN, serving 30% of all Internet traffic. With the broadest service portfolio in the industry, Akamai provides next generation CDN solutions for operators everywhere. Akamai Aura Licensed CDN is a suite of operator CDN solutions that enable next generation IP video services to deliver [sic] any device. Aura Managed CDN is a CDN hosting solution that offers a turnkey and highly scalable content delivery network with infrastructure managed by Akamai. And Akamai media delivery services satisfy audience demand for media streaming content and large file downloads delivered live and on-demand.

Yes, you read that right. Akamai CDN is not a consumer-focused CDN in its own right. Instead, it is the CDN for CDNs.

In fact, many names on our list—such as Fastly or StackPath, are based on the EDGE network, which in itself is part of the Akamai infrastructure. As such, Akamai is meant for CDN providers wherein they can build their setup atop it.

Homepage—https://www.akamai.com/uk/en/resources/cdn.jsp

CDN77
CDN77 claims to be the "Content Delivery Network chosen by Space Agencies." Naturally, it does have an impressive client base in that case.

---

[3] See: https://www.akamai.com/uk/en/resources/cdn.jsp Akamai

In fact, when it comes to web performance optimization—the core focus of this book—CDN77 seems to perfectly fit the definition.

It has two broad categories of offerings:

- First up, there is the static CDN network, ideally meant for websites and web projects such as web-based games and apps. The intent here is to boost performance by delivering assets via the global CDN architecture.

- The second option pertains to video processing and delivery, generally meant for video streaming and broadcast sites. CDN77 is not a minnow in the league of CDNs, and if the sole purpose of considering a CDN in itself is web speed optimization, CDN77 can be blindly picked and will work almost perfectly for most use-case scenarios.

Homepage—https://www.cdn77.com/

Netlify Edge

For most developers, Netlify is pretty much like a household name. With offerings such as free hosting with gracious bandwidth and resources for web projects, an open-source Netlify CMS that can act as a headless solution for app development, and various other useful things, Netlify has earned a reputation for itself.

Netlify Edge is a CDN that is more geared toward application hosting and delivery. Just like a standard CDN, Netlify Edge too supports static assets. However, beyond that, it also has several other features, namely:

- Netlify Edge can pre-render pages for faster performance

- It offers tight integration with Git

- It can be configured for auto-deployment and auto-rendering

Netlify Edge is ideal for developers looking to build web apps atop a JAMStack, as it integrates seamlessly with most other Netlify offerings.

Homepage—https://www.netlify.com/products/edge/

jsDelivr

jsDelivr is a free and open-source CDN with a difference. Instead of providing a wide variety of features such as DDoS protection etc, it provides mirrors for npm, GitHub, WordPress plugins, and other similar projects.

Now, it goes without saying that JavaScript powers the majority of the web. Combine it with GitHub, which hosts millions of open-source projects, and then WordPress, that powers more than 30% of the web.

jsDelivr offers mirrors for all of them, such that when you load a site powered by JavaScript or WordPress, you do not need to load resources from a geographically remote server.

- The usage is simple. For instance, to load a file from npm (in our case, a given jQuery file), we can simply use this code:
- https://cdn.jsdelivr.net/npm/jquery@3.2.1/dist/jquery.min.js
- Similarly, to load a GitHub release, commit, or branch, we will use:
- https://cdn.jsdelivr.net/gh/user/repo@version/file
- And lastly, for WordPress plugins, we can use the following to load one from the repository:
- https://cdn.jsdelivr.net/wp/plugins/project/tags/version/file
- The same structure works for WordPress themes as well:
- https://cdn.jsdelivr.net/wp/themes/project/version/file

jsDelivr has become a vital component of many developers' workflow, and owing to its open-source nature, its rate of adoption is rising with each passing day. It helps us use code and assets in our projects without having to worry about speed or downtime.

Homepage—https://www.jsdelivr.com/

Hostry

Hostry is a web hosting company that offers solutions such as domain names, VPS hosting, and other related stuff. It also has a CDN offering that is free for up to 10 GB of transfer and quite affordable thereafter.

Hostry CDN is not the most feature-rich one out there, but the fact that the initial data transfer is free can be appealing for some. That said, the CDN global coverage is satisfactory, but not the world's best out there. Here is the coverage map as of April 2020:[4]

If you are looking for a free CDN solution that can work well for specific regions, Hostry might well be worth a look.

Homepage—https://hostry.com/products/cdn/

---

[4] See: https://hostry.com/products/cdn/ Hostry

Swarmify

Swarmify is not a pure CDN per se. It is, instead, a video hosting and streaming solution that offers a mammoth set of features for video content. Such features include

- A custom video player, with customizable features

- Encoding support for various devices and platforms

- Tight integration with YouTube, including the ability to automatically fetch videos from it

However, other than all of the above, Swarmify also offers a video CDN on the side. This particular video CDN has a global network and can be used to stream and play videos seamlessly, with nearly no buffering and/ or delays. As such, it can significantly enhance the user experience on the websites that are hosting videos via Swarmify CDN.

Since video content is ever on the rise, Swarmify is likely to grow in stature over the course of time.

Homepage—https://swarmify.com/

Cloudinary

Cloudinary, much like Swarmify, is not a CDN service per se. Instead, it is more of a cloud storage and upload service that facilitates the usage and sharing of larger files on the web.

However, unlike Dropbox and other such solutions, Cloudinary allows us to upload and fetch assets to and from the cloud for use in our projects, be it a mobile app or a website. As such, we can store our videos and images and other such assets on Cloudinary, and then use Cloudinary's own tools to incorporate the same in our projects.

Cloudinary too has its own CDN offering for easier and faster processing of assets. However, it relies on various CDN providers such as Akamai, Fastly, and others for optimization and delivery to global locations.

Homepage—https://cloudinary.com/

That brings us to the end of this roundup of CDNs. At the end of the day, the basic question might arise—which one should we use?

The choice, broadly and largely, depends on our requirements and the nature of the project in question. For instance, a niche CDN solution such as Swarmify might be apt for a video-related project, or for showcasing and handling videos in our existing projects, but not for everything else. Similarly, if our target audience is in the Asia Pacific or China, a CDN such as Fastly or something of that nature might be a good fit.

On the other hand, if global optimization is our intention, perhaps nothing can come closer to beating Google Cloud CDN or Amazon CloudFront.

When it comes to speed and page load times vis-a-vis web performance optimization, CDNs play an essential role. As a result, we need to judiciously and smartly select the right CDN that fits our needs and requirements and can help us better optimize our web project.

Thereafter, we need to turn our attention to another equally essential and prominent aspect of speed-related optimization—that is, caching.

USING CACHING FOR WEB PERFORMANCE AND SPEED

A lot has been talked about the importance and role of cache on the web. In fact, caching in itself is not something unique to the web, and it finds applications in various paradigms of the computing industry.

In simplest terms, a cache is a static copy of a pre-existing dynamic piece of data. It stores data in a static format, such that any future requests for the said piece of data can be served faster.

In this sense, a cache does not really have to be a copy of a web page. It can be a software component, a hardware component, or both. In fact, caching in itself is a fairly diverse field of study that requires a good deal of time to gain expertise in.

Broadly speaking, it is safe to classify caching into four categories or "types," albeit this demarcation or classification is not fool-proof:

- Web Caching
- Data Caching
- Application Caching
- Distributed Caching

However, considering the fact that our focus in this book is practical insight into web performance optimization, we will obviously have to bypass all other forms of caching-related technical jargon and focus solely on web-related caching.

As a result, we will focus heavily and entirely on web performance-specific forms of caching and discuss ways in which using a good caching system can help us improve the overall speed and page loading times of our website or web application.

But before going any further, let us first discuss what caching actually does vis-a-vis web performance and speed optimization.

What Is Caching for the Web?

As we have already discussed, caching simply refers to the process of creating static and ready-to-serve copies of existing data in order to speed up response times.

This makes the definition fairly clearer—web caching refers to the process of serving static copies of website data or content in order to consume less time when rendering pages.

Now, since the goal is to improve the overall page load times, web caching in itself can be further subdivided into various categories. Of these, some specific ones such as proxy caching or gateway caching are less common than the other ones, and generally not required for most use-case scenarios.

That said, we will now be turning our attention to three specific forms of web caching methods and looking at each of them in detail:

Browser Cache

Browser Cache is a form of cache that is stored and used on the client-side, that is, the user's device. To be clear, Browser Cache is often viewed as a subset of Site Cache—the classification is not entirely unfair, as Browser Cache does work in assonance with Site Cache.

Nevertheless, the key difference is that the Browser Cache is stored within the user's web browser. As such, all cache files and contents are stored and saved locally, on the device itself, alongside other files of the browser. The web browser can then access them as and when necessary when browsing a given website.

Consider this example: we visit a given website today using the web browser of our choice. Thereafter, if we visit the same site later today or tomorrow or anytime soon, the web browser will be able to leverage its pre-existing cache to locally fetch certain page assets. This will imply that remote network requests for certain assets are not required—local files tend to load and open faster. As such, the website will perform significantly better, compared to a similar site that may not properly make use of Browser Cache.

This, obviously, depends a great deal on the web browser in question. Certain browsers tend to be better at local caching, whereas others might not do the job equally well. That said, most modern-day web browsers such as Chrome, Firefox or Safari are fairly good at browser caching.

Considering the fact that Browser Cache is stored locally on the user's device, the user can manually disable this feature, or delete the cache files as and when they want to. However, owing to good privacy controls offered by the modern web browsers, and also because it can significantly speed up the web performance, the majority of users tend to keep browser caching enabled.

Browser Cache works by means of efficient communication between the website and the web browser. As and when a page is updated, the cache becomes outdated, and the browser flushes it out and replaces it with newer cache files.

In general, Browser Cache is used to locally and temporarily store file assets such as HTML pages, CSS styles, JavaScript elements, images or thumbnails, and so on.

Site Cache

Also known as Page Cache, Site Cache is the most popular form of caching as far as front-end caching is concerned. Essentially, Site Cache creates temporary static copies of web pages, HTML and CSS assets, multimedia and images files, as well as other web elements whenever a page is accessed for the first time.

Considering the fact that Site Cache deals directly with the web pages and is stored on the server itself, albeit for faster loading, it is also broadly termed as HTTP Cache. However, in practical terms, HTTP Cache is a superset that combines both Page Cache and Browser Cache.

There is not much to understand as far as Site Caching is concerned, mainly because the core concept in itself is fairly simple and obvious. Consider this example: we might spend a while memorizing a few multiplication tables. In the beginning, it is a time-consuming learning curve. However, later down the road, certain calculations become faster and easier for our brain to process, as recollecting 5 times 12 is not a big deal for us.

Site Cache also works the same way. It creates easy to access copies of certain parts of the website data, such as certain elements of the page that need not be loaded again and again. Thereafter, for all repeat and future visitors, it serves the cached copy of pages in lieu of repeating the process of loading the pages from the scratch. This brings several obvious benefits to the fore—for one, the website in itself loads much faster and smoother. Other than that, database connections and server load is also reduced drastically.

Server Cache

Server Cache, as the name suggests, handles caching of websites and web pages on the server-side, not client side. As such, server caching is performed and the Server Cache itself is stored entirely on the web server itself, with the web browser or visitors having little role to play.

Server Cache is a broad field, and is generally heavily dependent on the server architecture in question, the tech stack of the project or website being hosted, and so on. Broadly, we can categorize Server Cache into sub-types of its own, such as:

- **CDN cache:** As we discussed earlier in this chapter, CDNstend to provide caching solutions of their own. Such solutions are not the most effective if used in solitary, but really prove useful if used in combination with other caching mechanisms.

- **Code cache:** Since most of JavaScript and other front-end scripts are optimized and cached by HTTP Cache, this particular form of code caching mostly deals with database-related back-end code, such as PHP optimization. Since PHP code needs to be compiled between various requests, it is a good idea to cache the same and serve cached copies of the code as and when possible.

- **Object cache:** Multiple database queries can slow down a website. To prevent this, Object Cache can be used to hold certain database query responses in the form of cache on the server, that can be served as and when requested. This can speed up database response times and also help in optimization of the database itself.

We will be turning toward Server Cache later on in the course of this book. Whilst we will look at the majority of server caching methods in Chapter 7, Object Cache for database will also be addressed in Chapter 6.

As such, for now, we will focus on HTTP Cache in itself, or Page/ Site Cache and Browser Cache. But before going there, let us spend a while enumerating why caching really matters for web performance optimization.

Importance of Web Caching for Speed

Fetching something over the network is both slow and expensive. Large responses require many roundtrips between the client and server, which delays when they are available and when the browser can process them, and also incurs data costs for the visitor. As a result, the ability to cache and reuse previously fetched resources is a critical aspect of optimizing for performance.

Google[5]

We have already covered what caching is, and the various types or forms of web caching. That said, what makes it really vital?

Here is a short list of the major benefits and advantages that caching has to offer, as far as web performance optimization is concerned:

Firstly, the biggest and most obvious benefit of web caching is that it helps improve site speed and page load times. Since most of the web applications and websites nowadays are dynamic in nature, they require multiple requests to the server and to the database, in order to facilitate flow of data. By serving cached static copies of certain pages or assets, the page load times are improved significantly as the number of requests and response times come down.

Second, caching can help in reducing the number of queries that are made to the database. Repeated queries can cause database overheads and, in case of websites such as forums or membership portals, the overall performance can slow down owing to transients in the database and overheads. Web cache can help reduce this burden by reducing the number of database queries as several requests are diverted to the cached copy of data and only the really essential ones are sent toward the database.

Third, web caching has several advantages of its own pertaining to SEO. We will be covering SEO in the next chapter, but for what it's worth,

---

[5] See: https://developers.google.com/web/fundamentals/performance/optimizing-content-effi-ciency/http-caching Ilya Grigorik, Google

caching has a very crucial role to play in SEO. faster websites get better rank in search results, period. By utilizing web caching, we can give our websites and web projects a significantly better page rank and SEO advantage over similar non-cached sites, all other factors of SEO being constant.

Lastly, web caching can lessen the server load. This is especially useful, as in case of CDNs, if we are using a shared hosting server or relying on a less powerful hardware setup. Since the cache is served from static copies of pages, database queries, and data access in itself is comparatively lesser. This implies there is lesser utilization of server memory and resources, and results in overall reduced server load.

As can be seen, web caching has enough and more positive aspects. This is why it has become a key component of web performance optimization, especially as far as speed optimization is concerned. No discussion of web speed can be complete without a mention of web caching.

With all of that out of the way, here is the big question—how should we implement web caching on our websites and projects?

Implementing Web Caching

Adding a caching mechanism to any given website depends on a variety of factors. Nowadays, the majority of content management systems tend to offer their own extensions or plugins for adding HTTP caching as well as other performance optimization methodologies.

That said, on a production server, it is a good idea to prefer HTTP headers for the purpose of caching, as compared to HTML meta tags. This does not imply that there is anything in particular wrong with HTML meta tags, but for what it's worth, many browser cache just do not adhere to them, and proxy caching does not read HTML meta tags at all.

As such, HTTP headers are the smarter way to implement caching. Here is what a typical HTTP header response might look like:

HTTP/1.1 200 OK

Date: Tue, Apr 21, 2020 13:19:41 GMT

Server: Apache/2.4.3 (Unix)

Cache-Control: max-age=3600, must-revalidate

Expires: Tue, May 26, 2020 13:19:41 GMT

In the above example, we are using certain Cache-Control responses within the HTTP header. The Cache-Control responses can be of varying nature, and here is a brief list of the same:

- **max-age:** specified in seconds, it denotes the maximum amount of time that a given cache will be considered useful or fresh.

- **s-maxage:** same as max-age but used only when working with Proxy Cache.

- **public:** all responses that do not require HTTP authentication are made publicly cacheable.

- **private:** this helps privately held caches to store the given response. A private cache is one that is exclusive to the given user (say, a browser cache stored on the device locally).

- **no-cache:** as the name suggests, it forces a fresh request to the server, instead of reflecting the earlier cached response. It is generally used on pages that require authentication.

- **no-store:** the cache copy is not stored.

- **must-revalidate:** it forces a revalidation on the current cache, as opposed to serving a previous version of the HTTP cache.

- **proxy-revalidate:** same as must-revalidate but used only in case of Proxy Cache.

We will be covering HTTP headers sent from the server at length in later chapters of this book when we address server-side caching measures. For now, it is worth noting that HTML meta tags are less preferable as compared to HTTP header responses for caching purposes.

Caching Operations

Generally, the majority of the caching operations are limited to GET responses at the most.[6]

Nonetheless, it is worth knowing some of the most common caching server responses:

---

[6] See: https://developer.mozilla.org/en-US/docs/Web/HTTP/Caching MDN

- 200 (OK) Response implies a successful request.

- 301 (Permanently Moved) implies that the given resource has moved to a different location.

- 302 (Temporarily Moved) denotes URL redirection to a new location.

- 404 (Not Found) error is fairly well known, and implies that the given resource was not found.

- 206 (Partial Consent) response denotes that the results being fetched are incomplete.

Useful Tips Related to Web Caching

Before we close this chapter, some handy tips and ideas:

- It is a very good idea to use consistent URLs for web caching purposes. Having the same content spread across inconsistent URLs only results in the same content being cached multiple times, thereby adding to the cache size.

- Servers should provide a validation token, especially in case of content that has changed recently.

- There might be certain resources that web browsers will already have stored in their browser cache—for example, Google Fonts or jQuery scripts. Similarly, certain assets might need to be cached by a CDN or likewise. As such, we need to be judicious when selecting which resources should be cached on the server, which ones to be cached locally by the browser, and so on.

- Depending on the nature of our site, certain cache might have a longer lifespan as compared to other cache files. For instance, sidebar widgets related to "About" or similar static pages may have a longer life span, whereas a "Recent Posts" widget on a news site or blog might need to have a very short lifespan.

- Using Pragma HTTP headers to force no-cache is a bad idea, in general. It is a common misbelief that adding Pragma: no-cache will imply in the given page not being cached. However, in reality, HTTP specifications do not have any practical or mandatory guidelines specified as of now related to Pragma response headers (only Pragma *request* headers are specified therein). As such, Pragma: no-cache might just be ignored by certain servers or caches.

- Much like a sitemap, website cache too can have its own hierarchy, making it easier for optimization purposes.

- When implementing caching on any site, churning needs to be considered. There are certain web resources that might be similar in nature, but be updated more frequently than the others—case in point, a highly popular JavaScript framework. To ensure that your code adheres to the latest updates and also that the cache does not need to be entirely revised every other day, such assets and resources can be placed in a different file and used/referred to therefrom. This will help us cache the rest of the content and code for faster results, and also lessen the amount of resources that might need to be frequently updated or downloaded remotely.

- It is not advisable to use POST methods far too often when working with HTTP cache. As stated earlier in this chapter, the bulk of the responses are in GET form, and more often than note, POST method responses are not cached.

- Always attempt to specify max-age and expiration-time with cache—this helps the cache mechanism recognize regularly updated pages.[7]

- Cookies, generally speaking, are tough to cache. As such, cookies should be used only as and when necessary, and not just for the sake of using them. Broadly put, static sites do not require cookies.

CONCLUSION

That brings us to the end of this chapter. In this chapter, we covered topics pertaining to web speed and performance.

Faster websites tend to do well not just in terms of user experience, but also in page rank and SEO, as we shall see in the coming chapter. As such, it is, indeed, important for us to consider speed as a key factor when working on the optimization of our web projects and websites.

---

[7] See: https://www.mnot.net/cache_docs/#TIPS Web.

Caching and CDNs can go a long way in helping us ensure just that. However, our discussion here on caching is slightly incomplete—no caching setup is ever complete unless it takes into account minification and server-side optimization. Both of these concepts will soon be covered in the coming chapters of this book, and that will enable us to have a clearer picture of web performance and speed optimization.

The next chapter will talk about SEO as well as localization or internationalization, from the perspective of web performance. As we discussed in this section itself, page speed does come into play as a factor for SEO. In the coming chapter, we will be discussing this as well as various other factors that can help us boost the page rank of our site whilst maintaining overall optimization measures.

CHAPTER 4

Conversion Rates: Localization and SEO-Specific Concerns, On-Page Optimization

IN THIS CHAPTER

➤ Learning about the importance of monitoring conversion rates

➤ Reviewing main principles of localization

➤ Analyzing mistakes to avoid when operating SEO

People spend more time on the digital world wide web in this day and age than ever before. They address all their needs and questions to online search engines and services that are only a few clicks away. Significantly more individuals opt for streamed social entertainment and online shopping rather than doing it offline. Regardless of location or the language is spoken, the web is growing exponentially to cater to everyone's wishes.

Businesses of all sorts were the first ones to seize the opportunity and utilize the web's potential, which resulted in an ever-intensifying competition for conversion rates. In the context of driven online marketing, popular economy consultant Peter Drucker said the following: "The aim of marketing is

DOI: 10.1201/9781003203735-4

to know and understand the customer so well the product or service fits him and sells itself."[1] For that purpose only, professionals build, test, and optimize their digitally exposed online enterprises. The idea of conversion helps them to measure potential success rate and impact of their web presence.

Here's how Google[2] defines conversion rate: "Conversion rates are calculated by simply taking the number of conversions and dividing that by the number of total ad interactions that can be tracked to a conversion during the same time period. For example, if you had 50 conversions from 1,000 interactions, your conversion rate would be 5%, since 50 ÷ 1,000 = 5%." To put it simply, it is the percentage of visitors to your site that complete a desired goal that could be a purchase, subscription, or consultation. Therefore, determining your conversion rate reflects the real picture of how well or not so well your website is doing.

You can use the above formula to calculate your website's overall conversion rate yourself, but in case you have multiple desired goals or indicators of successful operability, then you will have to track those separately from the overall conversion index. Shall you have any difficulties you can always outsource this calculation to online website conversion rate calculators that could complete it for you.

According to this US consumer report for 2020,[3] the average conversion rate across various industries is 3%. You can note what spheres have higher or lower conversion rate based on this breakdown:

Industry	Conversion Rate
Consumer electronics	1.4%
DIY & tools	1.7%
Automotive	2.2%
Home furnishing and decor	2.3%
Major chains	2.3%
Jewels and cosmetics	2.9%
Sports	3.1%
Others	3.4%
Apparel and footwear	4.2%
Health and pharmacy	4.6%
Gifts	4.9%

[1] https://marketinginsidergroup.com/strategy/marketing-is-business-the-wisdom-of-peter-drucker/, Marketing Insider Group

[2] https://support.google.com/google-ads/answer/2684489?hl=en#:~:text=Conversion%20rates%20are%20calculated%20by,50%20%C3%B7%201%2C000%20%3D%205%25., Google answers

[3] US Consumer Behavior, https://www.fisglobal.com/-/media/fisglobal/files/pdf/report/us-consumer-behavior-report-2020.pdf?la=en FISGlobal

However, it is worth mentioning that bigger e-commerce store that displays multiple products on the same website might have different individual rates for what presented above. In addition, rates can vary depending on what device you accessing the web from:

Device	Conversion Rate
Mobile	1.82%
Tablet	3.49%
Desktop	3.90%

It is rather surprising that website rates are still higher than mobile conversions; we tend to guess that it could be due to the fact that a potential customer gets twice as distracted with other applications and notifications when purchasing from smartphones and thus frequently decides to postpone it for later.

Now let's take a look at several factors and practices to apply that can positively affect your conversion rates.

- **Decide what you want to measure:** Before you get to know your current website conversion rates, you would be asked to decide what item you are going to use as a criterion for calculations. To help you figure that out, try to focus on what exact action you want viewers to take when browsing the site. Answers might include the following:

  - Sign up for updates

  - Sign up for free trial

  - Book a consultation

  - Purchase a service

  - Download an e-book

  Once you get more comfortable working with one of these items (or all of them) to track and optimize your conversion rate, you can get more advanced data on the percentage of new visitors or returning visitors, as well as the scale of organic traffic from different sources.

- **Make a personalized offer:** To attract as many customers as possible, you need to learn how to gather information about them through forms or cookies and offer exactly what they might want.

Personalization is a great tool that could be used to enhance your customer's service methods and deliver higher conversion rates. It aims at making your offer look irresistibly worth it for consumers. One of the best examples of personalization is free shipping. It is an indispensable and requisite practice in e-commerce stores that speaks to every individual that likes complimentary incentives. Thus, Walker Sands in The Future of Retail Report for 2018[4] stated that nearly 80% of consumers in the United States said to prefer an online store that offers free shipping over one that does not have that option. Additionally, around 54% of customers look for same-day shipping and are even willing to pay more for such service.

- **Practice remarketing:** It is a general understanding that very few customers make a purchase on their first visit to any e-store. So you need to gain a certain level of trust before buyers share their personal details with you. This is where remarketing comes in handy. It could be described as: "the technique revolves around marketing your products or services to users who have already interacted with you in the past; these can be your former clients or visitors who showed interest in what you have to offer."[5] It proves the fact that users are more likely to invest in a product that they already familiar with, or were interested in at some point. Thus, retargeting content and advertisement can definitely work and heighten conversion rates. To support the statement with some figures, approximately 25% of users enjoyed seeing remarketed content and as many as 43% stated that they were likely to convert.[6]

- **Upgrade customer support:** High conversion rate websites have one thing in common—impeccable customer support. Online service and support have a huge impact on the reputation as well as conversion rates of the website. Few things that can improve your performance are including FAQ on pages, highlighting guide remarks along with the site, and having an option of the live chatbox. The latest was said

[4] https://www.walkersands.com/wp-content/uploads/2018/07/Walker-Sands_2018-Future-of-Retail-Report.pdf, Walker Sands
[5] https://adoric.com/blog/what-is-a-good-conversion-rate-2020/, Adoric blog
[6] https://www.criteo.com/blog/power-of-retargeting-computing-high-tech/, Criteo blow

to increase revenue by up to 48% and boost conversion rate by up to 40%.[7]

There is a number of ways how to upgrade your current customer support quality:

1. **Be knowledgeable:** knowing your offerings in detail is essential for a successful customer service. Not only does it help with the delivery of your answers and gives more clarity, but it also allows you to be much more effective in communicating your answers which makes customers get the information they need faster.

2. **Be goal-orientated:** being goal-orientated completely changes how you engage with and interact with customers looking for help. It also gives your organization the perception of fast, helpful, and consistent service on top of an already great website offering.

3. **Quick responses:** almost everyone nowadays lacks patience. Being quick to return requests allows your viewers to know that you care about their issues and concerns, and are willing to help when they need you.

4. **Be personable:** being personable involves avoiding robotic responses and giving a personal touch when interacting with your customers.

5. **Be engaging:** make sure you follow up with customers a few days later to make sure that you are asking for feedback in regards to their perception of customer service they received.

- **Change and test again:** The only way to be sure whether certain feature works or not is to run A/B tests, as explained in the previous chapter. However, testing a high volume of content can take time. You should not expect immediate results in an increase of conversation but rather treat testing and optimization as necessary procedures that enable fail-free modus operandi of your website.

---

[7] https://www.superoffice.com/blog/live-chat-statistics/, Super office statistics

LOCALIZATION

A non-profit Globalisation and Localization Association (GALA) defines localization as the process of "adapting a product, an offering, or simply content to a specific locale or market."[8] Thus, it allows enterprises to take a chance and try succeeding in a different region and country from their home-based location. The decision to seek new markets may concern many businesses as it requires the obvious necessity to optimize much of their content in accordance to new cultural contexts and translating the whole website into other languages.

According to data from 2017 the Smart Insights e-commerce consultancy,[9] 62% of companies think of localization as too complicated and too problematic to occupy oneself with it. Nevertheless, it is good to keep in mind that not everyone who visits your site speaks your native language, which undoubtedly affects their decision to make a conversion. In that regard, presenting corresponding information about goods

---

[8] https://www.gala-global.org/knowledge-center/about-the-industry/language-services, GALA

[9] https://www.smartinsights.com/search-engine-optimisation-seo/multilingual-seo/7-trends-marketers-need-know-translation-localization-2018/, Smart Insight

and services to viewers in their mother tongues is more likely to result in improved website rates.

Localization means readjusting all of your content and not just the website. It should also include

- Service materials

- Product warranty records

- Product manual descriptions

- Online FAQ (frequently asked questions) forum and chats

- Personal disclosure declarations, such as terms and conditions

- Marketing materials, including the advertisement, social media posts, and videos

- Training materials for new branch staff recruitment

Starting from localizing the above-mentioned content would allow your enterprise to expand and be able to reach a completely new audience. Additionally, it would also help to restore loyalty and credibility among existing customers. Launching a localization campaign may look quite intimidating to some, but it should not be. Nowadays, there are many companies that would happily arrange all your necessary localization processes for you, given the set time frame and budget.

Ultimately, the most important factor that would make your localization project successful is your own team. When establishing a new base and working on delivering materials, the skills that your teammates possess would make a huge difference. They should be expected to perform great linguistic abilities (it should not be obvious the content you are displaying or the product you are presenting has been localized from another language) and technical quality assurance to be able to use up-to-date tools, secure and verify the localization is moving in the right direction. And if you are making enough efforts, local business partners and investors might notice your progress and approach you with collaboration opportunities that could be equally winning for both. In terms of enhancing visibility and monetization, you need to ensure your visibility by introducing more languages to your website and optimizing search engine optimization (SEO)–specific strategies for each expansion attempt.

It is a fact that localization has a lot of potential for brand expansion. However, you need to learn how to meet your intended audience half-way through offering them the content they want in the language they can understand. You have more chances of converting them into followers and customers if you go into localization carefully and gradually, thinking through every decision you take. In that aspect, opting for one-language-at-a-time localization instead of uploading multiple languages at once might be helpful. With that, you will be able to progressively expand your site's content and increase the SEO numbers organically. Few other things that one needs to consider in regards to language optimization include the following:

1. **Go with high-priority languages:** Once you decide on going world-wide and localizing your website in different regions, you may want to think twice about languages that you are going to introduce your business with. It is recommended to do an in-advance research study and watch the linguistic trends trying to see what tongue will bring more audience with it and therefore be crucial for expansion. For example, if you have a culinary blog, then localizing it for Italian or French viewers in their languages will bring more conversions than if you do that in Swedish or Kazakh ones. Knowing the specifics and main characteristics of the country you are aiming for would be helpful in this regard.

2. **Complete site-wide localization:** Partial localization is not considered to be effective compared to site and product-wide localization. Viewers who access your website from other countries should be getting the same quality treatment and level of curation as your home-based customers. Not delivering according to that principle would mean letting your audience down and failing expansion. To avoid that, you should be prepared to afford a well-arranged site-wide localization and translation of every piece of information on your web. If this seems to be a big bite to chew, you may consider outsourcing this part of optimization to other companies that can provide and publish your content in different languages in parallel.

3. **Language-specific SEO is the key:** In terms of processing search engine localization, it is important to remember that each and every language of your choice should be optimized individually. When it

comes to SEO, it's worth noting that all of your languages should be optimized individually. You can use online platforms like SEM Rush (https://www.semrush.com/) and Google Adwords (https://ads. google.com/home/) that have solutions for greater content optimization to determine the find out the course of action for SEO localization. Another great element to have that can boost your optimization efforts is adding a professional translator to your team. Having a native speaker will allow you to generate more original, country-specific content that has higher chances to show up on local search results pages.

4. **Keep in mind the text encoding:** Making sure the content management system you use reflects more language options should not be overlooked. It goes without saying that without correctly displayed text encoding other localization efforts might come across as irrelevant. Using Unicode Transformation Format (UTF) to encode various alphabets like Chinese, Russian, or Korean should be enabled on your site before additional language is embedded in its front-end content.

Mistakes to Avoid

Following these steps, the next thing would be to fine-tune your website-wide content to deliver the most personalized customer service ever. This would require many considerations of language, images, and cultural relevance. And in order to develop a strategy that surely reaches a global audience, we should familiarize ourselves with the list of the most common mistakes companies face on their way to successful localization.

Localization Does Not Mean Translation

In order to distinguish translation from localization, let's look at what each term implies. Translation, as described by Wikipedia, stands for: "communication of the meaning of a source-language text by means of an equivalent target-language text."[10] Localization alternatively includes conveying cultural meanings and bringing social and behavioral significances for the target audience to consider. Taking the process of localizing language just as the technical operation would be incorrect. Addressing your message inaccurately to the primary market might as well drastically affect your conversion rates.

---

[10] https://en.wikipedia.org/wiki/Translation, last edited on December 10, 2021

Insufficient Multimedia Localization

Just to be clear, it is not only the text, every image, audio, or video content on your website should be localized as well. Many sites rely on visual enhancements for advertisements like e-commerce companies and even personal blogs. Thus before you begin working on localization, you need to go through each piece of your multimedia and see how suitable it is for a new location (accept that some might need to be deleted due to cultural inappropriateness) or to what extent it could be adapted (if it is absolutely indispensable). It might be hard at this point to determine which photos and videos can stay and which have to go. Typically, it is recommended to get rid of the content that contains the following:

- Images or videos of famous people (politicians, influencers) in their professional setting could possibly alienate some of the new target audience that might not have a good association with the chosen figure.

- Animations or videos of impactful events (local audience might not feel the same importance of the event).

- The iconography that is just not universal across different cultures, therefore, should not be used to convey your message.

From a contextual perspective, it's also necessary that added multimedia content contains text descriptors next to it that can communicate the point you want to express with this particular image or video.

Lack of Data-Driven Analysis

Detailed research, market analysis, and thorough should be the core of every localization project. Before you set your foot in new territories, make sure you have completed SEO fact-finding that includes probing content plan, keyword research, and competitive analysis. With that, try to answer the following questions for each new country expansion:

- **Content Plan**
  - How does the audience here prefer to consume information? What formats do they like?
  - Do you have a content plan for each of these formats?

- What keyword combinations can be incorporated into newly optimized content? What keyword combinations should be avoided at all times?

- **Keyword Research**

  - What subjects matter to my audience?

  - How can we incorporate these interests in our content, products, and campaigns?

  - If we incorporate them, would it result in higher demand?

  - If we choose to ignore them, would it result in lower demand?

- **Competitive Analysis**

  - Who is your main competition for each new location?

  - What content format and keywords is your competition focusing on?

  - What relevant location-based practices can you learn from your competitor?

  - From which sites is your competition getting most of the backlinks and traffic?

  - What are the main local partners and stakeholders that your competitor has?

Lack of Culturally Sound Adaptation
The analysis above has to go hand-in-hand with cultural exploration that sets the tone in which you are expected to communicate to your viewers. Thus you might need to apply the following filters to guide you along the way:

- **Is the audience positions itself as culturally sensitive?**

  - What values and actions matter the most to the audience?

  - What emotions and actions are considered taboo to your audience?

- Will the tone expressed in original content translate well in the new language?

- Will the adapted content be as convincing as it is in the original language?

The list of these markers is endless and entirely depends on the destination country. However, the aim of this practice is the same for every place—you need to adjust your message to a new audience in an appropriate yet interesting manner to ensure a comfortable and seamless customer experience. Following this, it is also important to localize the language and the cultural relevance of your website. For example, two countries like the United Kingdom and Australia have the same language, yet the number of major and minor differences mean that you will have to modify and constantly regulate your spelling, cultural, and commercial approach to localized content.

Adaptation of Multilingual Service

For a business with ambitions, localization presents itself as a highly challenging necessity as customers' needs and expectations differ greatly depending on the region. However, localization is an overarching idea that goes beyond solving language barriers and deals with things such as cultural norms and nuances. So when we deeply contemplate it, we can consider localization to require all the marketing actions to correctly personalize offerings that bring conversion. It is also safe to say that poorly operated localization does more harm than good.

When thinking about localization, translation is probably the first thing that comes to mind. Providing content in users' first language is still the best way to boost the user experience. And while translation is not all there is about localization, it does play a significant role. It's essential to choose a translation method that strikes the right balance between accuracy and cost; but more importantly, suits the size, operation style, and focus of your website.

You usually have only two to choose from translating options: human translation and automated machine translation. When considering hiring professional translators, you can be sure for the quality of translations to be nearly perfect as they typically translate every material on an accurate page-by-page basis. However, some translators will not be able to multitask upload the content on the website; they would have to be working together with another team of engineers. Another thing is the cost of translation services. Professional service costs can get expensive, especially if you have more than a thousand pages to translate. And in case you do not possess that kind of budget, you need to think about other options.

On the other hand, automatic machine translation is regarded as a fast and cost-effective option when it comes to solving multilingual needs. Yet the matter of translation quality is often disputed when considering this option, even though we could recently observe online translations improving inaccuracy. Obviously, each translation method has its advantages and disadvantages to think about. The good news is that if you know your business processes well enough, you can combine both methods and choose one depending on the task you need to be completed. For example, if you installed online multilingual customer support on your site but may not be able to ensure the presence of support agents that speak a maximum number of languages at the same time, you can download automatic online chatbots. Chatbots could be a great alternative to professional experts and can communicate on your behalf in different languages maintaining the standard quality.

One thing you should not compromise is settling down for a mediocre translation technology. Different online translation solutions manage your content in different ways, and some of them are not viewed as applicable or advised practice for multi-language sites. Also, because search engines heavily penalize and rank low any duplicate pages and sites you want to avoid that practice no matter what. Instead, you need to place your localized content under the same Uniform Resource Locator (URL) as your original language within language-specific subdomains. With such architecture, duplicate content penalties could be avoided. If it confuses you much, online solutions like Weglot (https://weglot.com/) can automatically establish language-specific subdomains or subdirectories and also take care of tags.

To conclude, to offer good-quality content at all times, you need to find the right balance between automated and human translation or editing. In this way, you can have the best of both in terms of keeping the cost low and translation quality high.

Ignoring Design Considerations

Another localization mistake is not thinking thoroughly about the design of your website. It is important to recognize the role of design in the overall success of the campaign. Therefore, regardless of the content management system you are using, one of the basic things you should consider is creating a stylish, well-tailored theme for your site. The theme you choose has to use right-to-left formatting and be compatible with other plugins and applications. A well-structured theme even if it is customized, can incorporate all of the above elements and advance your site's functionality.

At most times, languages not only differ in syntax but also in terms of how much space their composition can take in the designated page area. Thus, when incorporating translated content into the design, you need to be careful not to miss the display and overall sense of the web front. Failure to foresee and prevent this could result in things such as overlapping text and broken sentences, which can hardly add any value to your website offerings. It may be concluded that you need to take website design into account and leave plenty of time to perfect your design to cater to dissimilarities and errors associated with localizing from one language to another.

Ignoring International SEO

When your optimized website is ready to launch for customers around the world, you surely want everybody to be able to search for it and eventually access it. This is where an international multilingual SEO strategy comes in handy.

International SEO basically means doing everything you do for domestic search engines for every new language version that you introduce to your site. Investing in your international SEO will make your site visible to customers around the globe that are searching in their own languages. Taking care of multilingual SEO involves managing language-specific directories, translating the wide-site content, and optimizing any metadata on your site including tags. If you have many languages to cover then multilingual SEO can look like a hefty business to run, but that is why you need a team of great enthusiasts and professionals that want the same thing you do—to succeed.

With that, localization is not just a campaign that has a completion date. You are required to constantly monitor your performance and optimize potential obstacles that come with operating in multiple languages. However, the resources you invest will definitely not be wasted as the role of localization in your conversion rates are pretty apparent. So you need to tackle localization with an open mind, understanding that it has great expansion opportunities to offer in return.

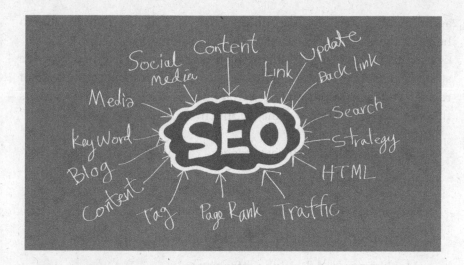

SEARCH ENGINE OPTIMIZATION (SEO)

It has already been a stated fact that you should consider localizing your SEO content in order to boost brand visibility on the Internet. In case you do not know where to start, it will be good to take a step back and observe the way people in your country of destination look for and interact with content. And if you are not familiar with international SEO rules at all, you can browse through this part to study several tips to increase your online traffic.

Create Audience-Oriented Content

If you want to take your business to another level with localization, then translating regular textual elements cannot be possible enough. Your SEO strategy should be focused on the creation of a deeper adaptation for your products and services. That implies becoming more acquainted with your intended audience, getting to know their Internet behavior and content preferences. Above all, your essential objective should be to provide targeted viewers with the most relevant and easy to access information.

When producing content, it is good to aim for something outstanding and attention-worthy. But with the volume of information that becomes available for viewers every hour, it could be hard to distinguish whether the content you publish will be perceived as notable. Naturally, you want to grow a wider audience by shaping offers to each market environment accordingly. Professional multilingual SEO specialists would be able to help you thoroughly localize your content to a different culture.

Another thing that can influence your multilingual SEO directly is the speed of your website. Making the website faster can affect the amount of regular and your ranking with search engines. Very little-effort actions that you can take to improve the loading time include:

- Optimizing the size of your images
- Enabling browser caching
- Installing a plugin that enables page caching
- Integrating a Content Delivery Network with your website
- Monitor your web hosting performance

Getting to upgrade the search results and rise in website ranking through localization is a complex task. To make it relatively easier you can start with testing and optimizing the website before you additionally create it in several other languages.

Keywords, Meta Description, and Titles

When creating a multilingual website, you need to monitor that it is not just the text that is translated, administrative, descriptive, and reference metadata is also crucial to gain more visibility and conversion rates. At the same time, what might be a winning keyword combination within the original version of your site, will not necessarily be accepted for your translated site. With that, prior keyword research of the new language is important to position right in the market.

A good starting point would be researching what keywords people use to lead them to your website. Once you list those words and phrases, you can now localize them to other target languages. However, since search techniques vary depending on the country, a primitive keyword translation will not deliver good results—you are expected to research key points for each new place. You want to look for combinations with a natural flow and be understood by local users and not just search engines. And since Google is penalizing "overstuffed" keywords and treats them as spam, the content you present has just the right amount of words that look appropriate to everyone.

The search for correct catchphrases can easily be completed by using online solutions that automatically detect your metadata and suggest a number of suited keywords for a particular country. Popular keyword explorers are Ahrefs (https://ahrefs.com/keywords-explorer) or Ubersuggest (https://neilpatel.com/ubersuggest/).

Well-written descriptions are also essential in the SEO localization process as they have to attract readers and make them want to click on your page. As mentioned in the previous chapter, you need to be careful with the length restrictions (not to make it too long) when scripting one for your website.

And lastly, make sure the title of your multilingual page, your text, or your campaign is interesting, inviting, and consistent with the context of other material. Localized titles should be brief but clear, giving away just enough information about what your web has to offer.

URLs

As a part of the overall presentation, URLs should not be ignored when localizing SEO content. Just like with keywords, it is important not to make URLs too long and incomprehensible. Another thing to avoid would be duplicating multilingual content. And while not all duplicate content is absolutely harmful, it could lead to penalties such as low ranking and indexing.

To prevent getting penalized, it is recommended to include language identification alongside the main URL. To illustrate, your original page could be www.example.com, while the English version could be that +the language ID—www.example.com/en/.

In case you are using URL with Country Code Top Level Domains (CCTLDs) structure, you can place the language indicator in dedicated subdomain and subdirectory as well. Thus your localized web can look in three different ways:

- Top-level domain (www.example.en)

- Subdomain (www.en.example.com)

- Subdirectory (www.example.com/en/)

Each of these is easy to set and maintain. Creating a unique URL and defining its linguistic URL parameters is important to avoid any SEO misinterpretations making it more compatible.

Additional specifics you should take care of and that can impact your rankings and visibility are the following:

- **URLs must be readable by everyone:** avoid generated URLs and customize them in order to provide full comprehension of your content. Both readers and search engines must be capable of reading your URL and get a clear understanding of what the page is going to be about.

- **Organize your content:** keep in mind need to categorize your URLs in order to clearly determine and show to search engines which content that should be prioritized over other ones.

- Do not use capital letter as it might confuse search engines and readers.

- **Prefer hyphens to underscores:** the way you separate words does matter. Google engines are set up to read hyphens and not underscores. If you want to get better ranking, make sure you apply this rule.
  Example: http://yourdomain.com/web-site and not http://yourdomain.com/we_site

- **Add your mobile URLs to a sitemap:** this way to can inform search engines that your web page is mobile-friendly.

- **Include your target keyword:** try your best to include your primary keyword in your URLs. And in case it cannot be done, attempt integrating your target keyword to other product or category pages.

- Block unsafe URLs with robots.txt or otherwise you can get penalized by search engines.

- **Canonicalize your URL:** pages can sometimes create duplicate content when they are dynamic pages with filters. You can prevent this action by using canonical URLs. This tag can be used if you want a specific URL to become the preferred one even if other ones direct to the same content. Example: <link rel="canonical" href="http://your-domain.com/web-site"/>

- **Don't forget 301 redirect for broken URLs:** in case you need to change a page URL, do not forget to inform search engines of its new location. You do not want to lose relevant link from a well-ranked page so that is why you need to implement 301 redirect on the old URL to notify Google bots your new URL destination.

ALT Tags

Since search engines still cannot crawl and decode pictures, developers use Alt tags to tell Google what the image is about. That is why it is important to localize them matching with the wide-site language and context. When thinking about what Alt tag to create keep in mind that it has to be:

- **Descript and specific:** Alt tags should describe the contents of an image in as much detail as possible. The more specific you can be when describing an image, the better.

- **Be relevant:** Alt tags should be used to describe exactly what an image shows.

- **Be unique:** Don't use your page's main target keyword as the Alt tag for every image on the page. Try to write unique alt text that describes the specific contents of the image, rather than repeating the contents of another.

Alongside keywords, meta description ad Alt tags, Google also uses hreflang attributes to help users determine both the language and the region it is aimed for. Complete hreflang definition is "an HTML attribute used to specify the language and geographical targeting of a webpage. If you have multiple versions of the same page in different languages, you can use the hreflang tag to tell search engines like Google about these variations. This helps them to serve the correct version to their users."[11]

Hreflang tags can be both places in the header section of the original page or inserted via a sitemap. For example, hreflang tag referencing English page localized for readers in the United Kingdom could look like this:

```
<link rel="alternate" hreflang="en-uk" href="http://
example.com/en/" />
```

---

[11] https://ahrefs.com/blog/hreflang-tags/, Ahrefs

Depending on how many regions the page is intended for, multiple hreflang items can be added.

When it comes to international SEO, hreflang tags offer additional value. They allow organizations to ensure that the site that is displayed to customers aligns with their needs and helps to create an ideal user experience. If you have different versions of your website for those who speak different languages or reside in different countries, verify that you have not made these common hreflang mistakes:

- **Not including the return hreflang tags correctly:** when one page links to another page as an alternate, that alternate must also link back to the original page. Failing to do so can cause errors in implementing the links.

- **Not using the default page:** setting the x-default option allows you to decide what these users should see.

- **Problems with pages the hreflang points to:** when your domain has changed, you have to make sure all your hreflang tags also receive the update. You do not want to have tags that point to missing or incorrect URLs or pages that return any kind of error.

One Language Per Page Rule

Translating the main areas of a page who; keeping the rest in original language could look like a tempting quick win option, but it really is not. Usually, web owners keep multiple languages per page in instances like forgetting to translate navigation text assuming users will understand it anyway; or leaving viewer-generated content like comments, reviews, and discussions in a different language as it is. However, multiple languages on-page could ruin the overall user experience. It could lead to readers not being able to understand the flow and navigation of the page or having trouble with comprehending the context or discourse over the site. Both situations more likely to result in unpleasant confusions and frustrations.

In this case, the issue could be resolved by adding the hreflang tags discussed earlier and translating all the material displayed on the page, combining automatically translated content with proofreading by a professional human translator.

In the next part of international SEO localization, we will discuss two of the most common threats (content and technical issues) that any developer or web owner should pay particular attention to when expanding your business.

CONTENT THREATS

Content implies much more than just the texts; it consists of images, videos, and clear structure. And if you find yourself contemplating on how your landing page should look and feel, you should start with a simple keyword research. The biggest content threat would therefore be not doing the keyword research correctly or skipping it altogether. Shall you commit this crime, you will miss out on key information about your targeted customers, including finding out how you can customize services and products for them. And even though it might be convenient to translate previously successful combinations on Google translate, in reality, completing decent research will get you far better results in terms of conversion rates and brand reputation.

One common mistake in creating a website in a different language is simply cloning the original version using an online translator. Many small and medium companies usually install translation software to save on translation costs in their content management system. However, in the end, it might not produce the content you were wishing for due to the poor quality of translated material. Another oversight would be not including native speakers from the beginning. Involving native speakers can ease the process of adjusting the main goals, messages, and strategies of an enterprise to reach local users incorrect, understandable way. One thing that could be done on a low budget but high quality is not launching the whole site at the start, but optimize more valuable, large margin content first, releasing the rest of the content gradually.

Localizing marketing content for your website is a crucial step to win over new users, gain their trust in your offer, and make them feel safe and home when they browse your website. Nevertheless, fully translating and localizing the display of your site is only one side of the coin. Another thing to take care of would be eliminating any technical threats that could harm your online operations.

TECHNICAL THREATS

The problem with technical SEO is usually not easy to track. Some errors that cause breakdowns or glitches are not obvious on the surface. The technical side is mostly about how to enable fail-free website crawlability and indexing. That is why the aspect of hreflang tag mechanism is important for SEO localization. As mentioned before, hreflang is operating to indicate the relationship between pages in different languages on your website to search engines. Hreflang attribute consists of three main components:

```
rel = "alternate" → suggests that there is an
alternative version of the website
href = "https://www.example.com" → this is the
absolute URL of the other page
hreflang = "en-US" → this part specifies the language
of the page, optionally with country code for the URL.
```

"Common hreflang mistakes" that most first-time developers make when localizing include:

- **Incorrect coding:** When designating hreflang attributes the ISO 639-1 language code is used as the main format. Usual error, therefore would be coding with incorrect initials. For example, using "uk" for the United Kingdom instead of "gb." Or "Es" for Estonia instead of the correct code, "ee."

- **Missing return links:** For hreflang to operate properly it is necessary that every line of hreflang code that references another page has the same hreflang code on every page it cites.

- **Incorrect or absent canonical tags:** Canonical tags "is a way of telling search engines that a specific URL represents the master copy of a page. Using the canonical tag prevents problems caused by identical or 'duplicate' content appearing on multiple URLs."[12] By using canonical tags, you let search engines know which page is your preferred page to index. If a canonical tag references to a different page or is omitted entirely, all of the versions of your page will be de-indexed.

- **Use of relative instead of absolute URLs:** Every link that is listed in the hreflang tags must be absolute URLs and not relative URLs.
  Example of absolute URL: https://example.com/example
  Example of relative URLs: /example.com/example

It is good to keep in mind when implementing hreflang that Google Search Console can report any errors that happen with your hreflang implementation. So if you are starting localization, this Google service will be helpful in terms of identifying any issues with your current back-end operations and fixing them.

There is also a number of other free and useful tools you can use for generating and optimizing your hreflang code on your page:

- **Hreflang checker and validator:** https://hreflangtest.com/

- **Sistrix hreflang checker:** https://app.sistrix.com/en/hreflang-validator

- **Aleyda Solis' hreflang generator:** https://www.aleydasolis.com/english/international-seo-tools/hreflang-tags-generator/

When creating a localized site, there are many SEO factors to think about. For instance, translating and adjusting your content, going through

---

[12] https://moz.com/learn/seo/canonicalization#:~:text=A%20canonical%20tag%20(aka%20%22rel,content%20appearing%20on%20multiple%20URLs,. Moz

keyword research, and checking your hreflang code. Yet it is good to remember, that addressing these issues, if done in a proper and scrupulous manner, will only boost your website rating and bring more profit.

CONCLUSION

Before going international, one must identify own resources and capacities because expansion in many ways does not differ from creating a business from scratch. Making sure you are aware of all the implications and ready for additional costs that cannot be avoided, like translation or metalinguistic SEO could be a good foundation for localization strategy. If you are hoping to be able to make it without one, you will soon realize that reaching targeted sales without needed investment would be nearly impossible. Global SEO requires not only financial resources but also collaborative effort across keyword, cultural research, and creating local initiatives. And when we localize, the amount of previously mentioned things to consider can often seem like a complicated perspective for any business seeking growth. Luckily, effective international SEO can ensure brand consistency worldwide as well as maintenance of local performance. With that, there are also many great online tools that can facilitate localization and help optimize your website at lesser costs.

CHAPTER 5

Measurement and Analysis: Analytics Tools and Measuring Web Performance

IN THIS CHAPTER

➤ Reviewing the necessity of web performance measurement

➤ Choosing the right web analytics tool

➤ Studying how to interpret analytics data

If you own an online business, you will always compete with other e-commerce sites for consumers' attention. And having access to your marketing indicators and data will make it easier for you to generate the right content and send a targeted message for consumers, utilizing your channel and marketing efforts to the fullest. In the long run, this will result in higher conversions and a stable return on investment. The key feature of any professional marketing research is web analysis, which consists of collecting, measuring, and interpreting data about your website. Among many other things, web analytics can allow you to understand how viewers engage with your website—how many users are you

hosting per hour, what pages attract them the most, and how much time they usually spend browsing it.

Web analytics is concerned with tracking and monitoring metrics as such to measure your website's performance and update some items to bring a more positive user experience (UX). Ideally, you are expected to use analytic insights to optimize your marketing strategies and develop the right combination of operating methods that would complement customers' habits and preferences. It is indeed considered a great tool to boost your online visibility and tailor users' experience to improve conversion rates.

Web analytics might seem like a complex practice for someone who is just setting out, yet fortunately, there are plenty of useful and easy-to-handle software services that will make it manageable for you. Reports from these analytics tools allow developers and marketing experts to understand customers' interactions using the following types of raw data provided in web analytics reports:

- **Web traffic:** the group of incoming and outgoing website visitors within a certain time-frame

- **Direct traffic:** the number of viewers that accessed your website by going to your address directly, without making use of a search engine

- **Organic traffic:** the number of viewers that access your website directly from a search engine and not from other social media channels

- **Unique visitors:** the number of first-time visitors

- **Clicks:** the number of times a link has been clicked

- **Views:** the number of times a page has been browsed

- **Conversion rate:** the rate of viewers completing the desired action at your website (purchasing, signing up for newsletters, or subscribing)

- **Bounce rate:** the percentage of viewers that leave the site instantly without interacting with the website

Web analytics allow webmasters and marketers to see a visualized map of user interaction with the website. In simplest terms, the data report can present numbers of visits or clicks, but beyond that, there is also a possibility to make use of metrics to a greater extent and obtain information that would help to convert viewers to active customers. Ultimately, by studying behavior maps, an expert can tell what pages are relatively effective and which areas are causing confusion and distractions for the viewer. A good data manager can study data sources, examine various output ranges and set key performance indicators (KPI) or benchmarks to measure the attainment level of the website. Usually, websites have more than one KPI, and each one of the indicators should be monitored and reviewed separately. The way one chooses analytics tool majorly depend on the size of market reach, the purpose of analysis and KPI demands. At most times, companies with large digital outreach choose to outsource web performance analysis to professional data agencies.

Most services present their research on customer insight and behavior in detailed graphs and visualizations of the data in real-time. Additionally, web analytics agencies can also monitor the status of your online presence in social media posts and shares. They also might be able to give you a detailed assessment of different traffic volumes coming from several social media channels and tell you what decisions should be made for better site quality. Moreover, the service can help you monitor A/B testing to enhance the web performance and implement permanent changes to the web structure or content.

Every website is likely to have a set of its own goals and intentions, and each and every goal has to have a separate set of objectives or KPIs that outline a course of action that is necessary to take to achieve the overarching goals. For instance, an e-commerce website may need to measure more than ten different metrics to assess the performance of their platform—sales, bounce rate, sign up rate, leads, percentage of new visitors, traffic volume, unique visitors, and other KPI types. For those who want to experiment and test some feature or introduce new content, A/B testing tools or multivariate testing tools would suit the most. And enterprises that need market research should be using keyword research tools and competitive analysis tools. Enterprises that need market research should be using keyword research tools and competitive analysis tools.

And while most web managers prefer to stay away from selecting and using more than one analytics tool, it is certainly necessary because each tool has been masterly optimized to test the different criteria of a site. And since operating multiple tools for monitoring the performance of KPI scores seems daunting for many, it is one of the main reasons why website analytics service is still in high demand. Professionals with experience are likely to already have all the needed tools for systemic analysis and are therefore expected to deliver more accurate results and assist to uplift the return of investment rate.

WEBSITE ANALYTICS SERVICE BEST PRACTICES

Website analytics can strongly reinforce both qualitative research and quantitative results when reviewing the KPIs of your website. It is important for the analysis tool or testing solution you pick to conform to some of the core practices considered highly efficient in the field. Few best practices to keep in mind include the following:

- **Avoid only statistical reports:** New website owners may usually be misled into thinking that reports consisting of only numbers of visits, various rates, and page views are enough. However, reporting only large numbers could be deceiving and does not provide a full context of site performance; just because there is more time spent on site and fewer bounce incidents does not necessarily mean progress. Hence, superficial reports containing only data should be discouraged; what you want to see is an in-depth analysis that can give you something specific to learn about what is behind the data provided and what needs to be done to attract more organic traffic.

- **Always provide data with insights:** In addition to the previous point, reporting data without expert insights will be qualified as incompetent regardless of its accuracy. Every report should be coupled with some meaningful information regarding how the data shows areas of success or points where you lose most traffic and what immediate solutions could be applied.

- **Encourage data-driven decision-making:** The website analytics tool you opt for should tell you whether you have met or failed to meet your goals and help you work on how to improve your KPIs. A website analytics expert must be able to reveal if your site has irrelevant content that is not getting enough traffic, or how you can advance the display of the page to improve viewers' perceptibility. Find out how you can leverage testing tools and decide on the best arrangements that generate the most engagement with the page through user or engagement analysis. Therefore, it is important to communicate with the tech team and directly involve them in the decision-making process of improving web traffic and conversion flows.

- **Avoid being snapshot-focused:** Paying attention only to the rate of visits or looking only within a specific time period does not represent the wider and more compound web occurrences that are happening online. Other metrics that can provide a longer term understanding of users like pan-session metrics or user-lifetime value allow you to evaluate how the website evolves over time as it grows a new audience and interacts with returning visitors.

- **Communicate clearly with stakeholders:** The information you provide to your stakeholders should be consistent and clearly disclose your website performance's strengths and weaknesses.

Working with website analytics can reduce the complexities of running your business and provide explicit reports you need to be acting upon to generate tangible impact. The information your team should receive after your website was analyzed should preferably be:

- **Search-focused:** data should be designed to offer insights and be of maximum use for your specific intended marketing campaigns or search engine strategies.

- **Offering only essential data:** operations' dashboard should not be full of confusing graphs or charts for your team members to struggle to untangle it. Web analytics should enable you to construct a comprehensible interrelationship between analysis and action, that could later provide you with a candid view of the key statistics and subsequently determine the success of your search marketing campaign.

- **Integrating web analytics information into an actionable platform:** analysis and reporting are only one part of the process, acting on the insights acquired from this data is just as essential. Collected data should enable you to take quicker actions—improve advertisement content, post upgraded visuals, or introduce new keywords. As soon as the analytics tool has gathered the information for you, it needs to be responded to and implemented. If you cannot actually act on your data, then web analysis has no value to your business whatsoever.

A marketing research-focused solution, should not just offer raw data leaving you to interpret it for yourself. Analytics should be integrated into a shared workspace (without overflowing) and drive the updating and expanding progress of your web. What is also important about the analytics data is that it is fully actionable from the moment it is gathered and interpreted. Ideally, integrating web analytics into main business processes should enable you to:

- **Be more efficient:** being able to see which actions and display of the content are driving or hindering traffic flow should enable you to better prioritize your assignments and efforts in marketing and search engine optimization (SEO) campaigns.

- **Improve your results:** in order to effectively capitalize on the rationalized set of metrics with web analytics, one has to have a means of actually acting on the data being presented. Being able to review and react accordingly to the critical directions is essential to creating successful strategies.

Do I Need Web Analytics for My Website at All?

If managing a website is your business or generates a portion of revenue for your offline business, you definitely need website analytics through online software or professional services. It is crucially important to understand customers visiting your website, see how they use it, and find out why some of them might leave your site empty-handed. Just like you would require market research to understand why customers bounce off your business without making a purchase, you need to know what your customers have to say about their experience in general. Their feedbacks and online comments can be more profound and helpful than you can imagine.

Websites that primarily deal with e-commerce will have the most benefits from using web analytics software. Website performance research can give each website owner or developer the chance to monitor their customers, improve some trouble settings on their website, and focus on improving the conversion rate of new customers. Their UX can be refined by observing bounce rates and volumes of organic traffic using various web analytics software. Nevertheless, even small websites can benefit from seeing visualized marketing data. If you have initially invested in any website business, it is only sensible to want to know how well your investment is carrying out.

TYPES OF WEBSITE ANALYTICS

Website analytics is fundamental for every website. Investing in web analysis can highlight the factors responsible for low traffic, and tell you exactly what adjustments you might need to administer to improve its conducting. It can also tell you when Google is introducing new updates and notify what metrics you should pay attention to for better ranking.

In many ways, your website serves as the focal point of all marketing efforts. It can be used to share the brand's history, promote certain products or services as well as directly communicate with customers, and more. Web analytics tools can help you resolve nearly any concerns you have in regards to your website or customers. But there are a lot of tools to choose from today, here you can learn about five types of web analytics, what each of those tracks, and how they can be used to improve your online marketing endeavors.

User Analytics

It goes without saying that this type of analytics has a lot to do with finding out who your visitors are and what can be done to deliver the best UX to them. If you request the data about their average age, access location, or gender, it would be much easier for you to integrate items that could help them navigate and enjoy your website more. For instance, if you know where most of your users come from geographically, you can prevent any potential problems with language usage. By obtaining this data, you avoid unpleasant miscommunication experiences and provide an alternative option to using content translated in their own language. In addition, this type of analytics can tell you the type of device that customers use to search and access your site. In case most of them browsing through mobile devices, it might be more profitable for you to start optimization from mobile version or mobile advertising.

Content Analytics

With this analysis, you can specify what kind of visitors are engaging with your content: whether they are satisfied with just the average materials that you present or they looking for more technical information for which they are going elsewhere. You can as well identify what is your low-performing and high-performing content. The most favorable content would be getting more mentions, likes, and shares; materials that do not get as much appraisal should be analyzed in terms of how their keywords, headlines, and design layouts might affect the lack of engagement. Later on, these insights can help you to create tailored or slightly enhanced content to target your market and drive a higher audience to your site.

Behavioral Analytics

Every visitor once entering your site leaves a specific digital footprint that you can look into to see whether they have some behavioral patterns in common. Behavioral analytics can offer an opportunity to see what people are actually using your website for and on which pages they tend to spend the most or least of their browsing time. Having this data, you can monitor which pages are performing well and which should be optimized to become more attractive. The main focus should be to observe how users interact with the website from the moment they access it until they decided to act and buy your product or sign up for a subscription and more.

With that said, the end goal of behavioral analytics could be defined as a tool that can offer a seamless and untroubled website experience with fast-loading pages, clear design, and a user-friendly interface.

Traffic Analytics

In case you are running a SEO campaign, then you might want to know what website or platform directed the most traffic to your website. With traffic analytics, you can get that data instantly and start making more marketing efforts in that direction. Nowadays, the most directed traffic comes from social media platforms. It is a common practice that someone might come across your brand on Facebook or Instagram and therefore decides to visit your site. If traffic analysis proves for your case to have similar channels, you need to make sure to leverage the social marketing department forward.

Acquisition Analytics

Unlike previously mentioned traffic analytics—acquisition analytics focuses not only on what are the main sources of traffic but tells you whether those sources are actually helping you to acquire leads or lock new customers. If visitors that are redirected from other sites end up requesting your service or installing your solution then it could be considered as a good indicator for further partnership with those sites; but if the directed volume is not generating any revenue for the business, you might need to reexamine your marketing strategy so you do not misallocate resources on unnecessary campaigns that do not produce results.

To conclude, you have to keep in mind that you can find all the information about your customers and web performance indicators you will ever need underneath your website. And taking into consideration the rate at which data is growing, you need to be proactive and take charge in monetizing the value of data for your business. To make this happen, you should analyze your data by incorporating web analytics in your management style.

Social Media Analytics

Nowadays it is impossible to do business without considering social media. That's why it is important to analyze what happens with your content when it lands on social media accounts. You need to be able to see not only the headline figures on how many times your content has been shared, but it is also necessary to assess interaction, engagement, conversation, reach, and others. When considering social media analytics, you should be looking at the following indicators:

- **Number of Followers:** is a good indicator of potential reach, and similar to your email list size, you want the number of followers to grow at a steady rate.

- **Engagement:** is a reflection of your influence on social media. You want to monitor whether people liking, commenting, or sharing your content. As well as pay attention to what types of social media gets the most response.

- **Traffic:** you can see if viewers are using the links in your social media to visit your website.

- **Reputation:** new social media channels are great platform to find out how others perceive your business and respond accordingly.

WEB ANALYTICS PLATFORMS

By applying one of the 18 shortlisted advanced web analytics platforms available online, you can now make more informed decisions to raise your website's efficiency and conversion rates. It is as easy as ever to address the assessment of your site to a range of services that track data and gather research reports for free.

1. **Google Analytics:** Google Analytics is one of the most popular analytics tools out there at the moment. It is a Google service that lets you track websites, social networks, blogs and provides customizable business reports as a result. In simple terms, the way Google Analytics operates is that it inserts tracking code into the code of your website. With that, various activities of your users and their attributes (age, gender, interests) can be easily tracked. Collected data then can be combined into four main levels: user level (number of actions by each user), session-level (actions per individual visit),

page view level (number of pages visited), and event-level (number of clicks, comments, etc).

Main features include

- Google Analytics is created to organize your data in a more efficient way

- This website analytics tool supports both desktop and mobile applications

- It gives you insights into how people use your website

- You can choose from a variety of reporting tools

- Google Analytics can be integrated to work together with other Google products

Homepage—https://analytics.google.com/analytics/web/provision/#/provision

2. **Mixpanel:** Mixpanel is considered the best analytics tool to monitor KPIs progression and track user behavior with mobile and website applications. Mixpanel also offers data integration and data management services, and in case you need a platform that can be used as a team dashboard too, Mixpanel would be an ideal option.

Main features include but not limited to:

- Mixpanel allows you to visualize how users browse your websites

- You can divide viewers into different groups and compare them

- You can divide your data into separate sections for better visualization

- It allows you to create and keep user profiles

- You can monitor and analyze your website conversation rates.

Homepage—https://mixpanel.com/behavioral-analytics/

3. **Crazy Egg:** Crazy Egg is an online application that has many features to track user behavior on your website. It can also be used to boost your site profile and enhance customer interests through analysis and fine-tuning.

The main features are

- Seeing how people interact with your pages

- Crazy Egg distinguishes between active users, new users, and users using a mobile application

- It allows you to track what people are doing with available forms and login sections

Homepage—https://www.crazyegg.com/

4. **Heap:** Heap is software that has an easy-to-use interface and defines itself as "an analytics platform that helps product, marketing, and customer success teams craft exceptional digital experiences that convert and retain users. We give you all of your customer data—automatically—and the tools to turn that data into action."[1]
   Main characteristics include:

- Providing detailed reports that give context to data

- Allows tracking product and user engagement

- It can be a great platform to reconstruct a more engaging UX

- It allows you to measure website conversion rates

Homepage—https://heap.io/

5. **Motomo:** Motomo is an open-source web analytics tool that encourages you to have "100% data ownership." The company positions itself as the Google Analytics alternative and provides flexible features with absolute privacy protection.
   Features include

- The tool allows you to track and take control of your website data

- This app enables you to relocate your data between different hosting options

- You can see viewers' profiles

---

[1] https://heap.io/, Heap

- You can customize application features and icons the way you wish

- Installed in General Data Protection Regulation (GDPR) solution

Homepage—https://matomo.org/

6. **StatCounter:** StatCounter is an analysis website mainly focused on web traffic. It helps you to track and analyze your visitors' activities and responses. You can work on increasing traffic and SEO techniques more efficiently if making use of such detailed metrics estimates.

Popular features include

- Examining your website traffic trends

- Providing a full picture of users' visits to your website

- Setting alerts to know when an active visitor returns to your website

- Monitoring how many times a website user is clicking on links or advertisements

- Available information about visitors' operating systems and screen resolutions

- Helps you to identify the most visited pages

Homepage—https://statcounter.com/features/

7. **Adobe Analytics:** Adobe Analytics is a great tool to discover and measure the technical dimensions of your website. It enables data management from any sources such as mobile applications and various websites.

The main features are

- It can collect and process data from any source

- It offers better insights from using machine learning and artificial intelligence

- Can search for unexpected patterns of website performance

- It offers detailed and apprehensible reporting

- You can create all sort of additional data alerts

Homepage—https://business.adobe.com/products/analytics/adobe-analytics.html

8. **Finteza:** Finteza is another advanced analytics tool that mostly tracks traffic, and then advertising and conversion. This application allows you to compare the performance of various web pages as it supports more than 50 System Development Kits and Content Management Systems.
   General features include

- It can help you manage advertising banners and edit landing pages

- Detecting low-quality traffic sources

- Filtering out various pages and keywords

- The tool provides real-time conversion rate calculation

Homepage—https://www.finteza.com/

9. **GoSquared:** GoSquared is a real-time web marketing analytics tool that helps to enhance your website conversion. It also offers beneficial options like a team dashboard and managing to work and access data from different devices.
   Common features include

- It provides an accurate live measurement of how many visitors are currently browsing your website

- You can see what languages visitors opt for on your website

- You can view operating systems used by the visitors on your website

- It can send traffic-related updates via email

- It can estimate and report your daily, weekly, or monthly traffic

- You can discover whether website visitors are browsing from their desktop, mobile, or tablet devices

Homepage—https://www.gosquared.com/

10. **Hubspot:** HubSpot is a "platform that has all the tools and integrations you need for marketing, sales, content management, and customer service."[2] It enables your team to monitor leads, traffic, and other key indicators of your website in one place.
    Main characteristics include

    - It provides web pages, landing pages, and blog templates

    - It supports Office 365 and Google calendar

    - 24/7 live chat is available on the website

    Homepage—https://www.hubspot.com/

11. **Clicky:** Clicky is an online application that not only allows you to view the number of visitors on your website but also if any users are clicking on any of the links that point to third-party websites.
    Main features:

    - It enables you to observe your website traffic without leaving the site

    - You can view the total number of pages visited by users

    Homepage—https://clicky.com/

12. **Internet:** Internet is another web analytics software that can analyze not only your web but also mobile traffic. It has numerous features to offer and to support you in making a more effective decision based on user behavior.
    Features:

    - Can automatically identify any breaks in web conversion rate

    - Keeps you on track of any news events as well as technical website crashes

    - Offers you a platform to examine UX

    Homepage—https://www.atinternet.com/en/

---

[2] https://www.hubspot.com/#, Hubspot

13. **W3counter:** W3counter is a tool that is mainly focused on user behavior. It allows you to learn more about your visitors through simple reporting. Additionally, you can track your traffic online and see the exact sources.

    General characteristics include

    - You visitors' access locations and languages that they choose

    - You can find out what devices visitors use to browse your website

    - An easy-to-use application that requires zero coding background

    Homepage—https://www.w3counter.com/

14. **Gauges:** Gauges is one of the best web software that provides real-time site analytics. You can monitor your traffic status online and share any statistics with your team via Gauges Share option.

    Features:

    - The application also provides a live map that is updated when anyone visits your website.

    - You can get page by page detailed traffic analytics.

    Homepage—https://get.gaug.es/website-analytics/

15. **eTracker:** eTracker is an analytics platform that assists you in tracking users' visits. It provides end-to-end solutions and allows you to request reports compiled the way you like.

    Popular features are

    - Monitoring which products are viewed frequently

    - It enables you to observe items in the shopping basket

    - Helps you to track users' scrolling patterns

    - It is suitable for both web analytics and application analytics

    Homepage—https://www.etracker.com/

16. **Woopra:** Woopra helps you to identify who your customers are. The tool also allows you to calculate your product consumption by

analyzing web campaign performance. It is one of the visitor tracking tools that offer visualized reports.

Key features include

- Measuring clicks and conversion rates

- Observing how users navigate your website

- Helps to compare various user behavior segmentation by geography, date, and industry

Homepage—https://www.woopra.com/

17. **Coremetrics:** Coremetrics is a web tool by IBM that provides an online service of real-time analysis and data interpretation of your website. The platform has a dashboard that could be made accessible for your team and supports more insightful teamwork.

Coremetrics offers

- You can create, revise and share your analytics with others

- You can choose to implement free tools to increase your website conversation rate

- Enables both customer behavior analytics and cross-channel users' analytics

Homepage—https://www.ibm.com/digital-marketing/coremetrics-software

18. **Webtrends:** Webtrends is a digital analytics tool that offers to collect main website metrics and provide data on new and returning users of your website.

Key features include

- You can customize the team dashboard in any form you like

- Reports can be downloaded in PDF files

- The tool identifies how users access and share use website content

- Allows you to track clicks and review your website navigation effectiveness

Homepage—https://www.webtrends.com/

As you can see, there are a lot of web analytics tools—and the information that each provides is very attractive. You may think that adding more tools is going to make monitoring more efficient. However, doing that you might fact tool fatigue. Tool fatigue results from usefulness of the same tools because each of them adds a cost—in the form of time, money, and your attention. So how can you find out which web analytics tools are best for you? The short answer would be to see your team's priorities.

HOW DO I CHOOSE THE RIGHT WEB ANALYTICS TOOL FOR MY WEBSITE?

It is important for an enterprise of any kind and size to engage in scrupulous and extensive research before committing to a particular web analytics solution. Some of them could be harder to install while maintaining others may demand special skills from your team. The process of decision-making can be time-consuming and depend only on each businesses' key requirements and limitations. For instance, state agencies tend to have rather standard demands for data analytics but very strict legal and procurement-related purchasing constraints. On the other hand, tech enterprises usually have uncomplicated restrictions but complex and specific needs. In the end, every company has to make its own decision. The following 11-point survey for comparing analytics tools might be helpful when making up your mind:

Start by Drafting a Solution Features Wish List

Deciding what analytics solution to opt for should start with a thorough team brainstorming session. The team has to consist of experts from different fields such as marketing, engineering, support, and research departments. In such a multi-functional pot, everyone has to contribute and present professional opinions as to which platform would serve their objectives the best. Before commencing the evaluation process, it is recommended to create a list of necessary features that the application has to own. Which items would be ranked as crucial and which are viewed as additional? Talking through and compiling such a checklist would allow you to see what cannot be compromised and afterward compare different options empirically.

Examples for the general feature list might include

- Messaging and A/B testing

- Marketing research feature

- Can track users across desktop and mobile applications

- Offers visualized reports

- Detects errors and alerts instantly

- Offers operational support

Define Your Targets

The team is also expected to establish what events do they want to track with the analytics software. Begin with determining what are your high-level goals such as revenue rate or market share, and then spell out KPIs and metrics that imply progress toward main objectives. With that, you need to pick out the platform that has all the necessary aspects to deliver and track outlined KPIs and web performance metrics.

Analytics Data

The most important priority should be the software's quality of processing user data. Therefore, it is necessary to find out the platform syncing mode or whether it provides a real-time monitoring option and how reliable it really is. And because not all the parameters are usually included in the application documentation, it is essential to understand what might be its hidden features or limitations. Potential deal-breakers for tech enterprises may include segmented data by a fixed number of fields or over a certain time period.

Technical Requirements

The analytics platform that you choose should be consistent and trustworthy. Every company has its own information framework, and the solution has to be compatible with any data format, size, or limitations that you might be facing. And if there are more requirements in regards to hardware, CMS, or multimedia, you have to make sure it will not interfere with the overall analytics system.

Legal Requirements

Before signing any user agreements, you need to make sure that your country's data protection laws do not contradict with analytics provider's terms and conditions of service. Scrupulously reviewing legal requirements might prevent you from accidentally violating any laws and protect your interests as a customer. And surely, before settling for analytics outsourcing, you must review the provider's terms of service to confirm that your team is indeed the one keeping hold of all the data ownership.

It is also recommended to consider any long-term risks that might happen in case your provider goes bankrupt—will your data still be safe and accessible? To answer that, you might want to do a certain background check on the company's financial conditions and history of previous performance in business.

Design and Export Capabilities

If you are going to interact with the analytics platform often, you need to see whether its interface and display design suit everyone from your team. A good design would be the one that helps experts to find tools and answers faster. Practical and well-thought even in terms of aesthetics solution is more likely to bring a higher return on the project investment. On the other hand, a disorganized and simply lousy interface will be left unutilized.

Furthermore, some solution brands are known for making data exportation to other applications difficult on purpose. A good analytics platform would be the one that does not make you dependent solely on its functions; it lets you export the data to your own database if needed.

Professional Support

It is the professional support that analytics companies usually send the most mixed messages about. Claiming to provide too much support might suggest that their product is high maintenance. Yet saying that no support will be required may indicate that the software lacks resourcefulness and adaptability. You should be looking for a middle ground: a multifunctional platform with just enough support to operate with it.

Another thing to consider is how much time your engineering team might need to implement the platform. If the integration is too complex and time-consuming, will your provider be willing to offer 24/7 and is that going to cost extra? If not, then investing in additional staff training, onboarding, and maintenance may require additional funds.

Integrations

In order to consolidate into your pre-built data setting smoothly, you have to check how the solution is going to integrate with common platforms such as Customer relationship management or Enterprise Resource Planning, and other support tools. Can it assimilate with your website performance dynamics? Ultimately, you want to look for an extensible and multifaceted platform that will grow more helpful over time and adapt well to your team structure.

Demonstration or Trial

Before purchasing an analytics solution, it is preferable to verify the platform's operational capacity with a trial. If it is possible a cross-functional planning team should be gathered for a demonstration and then a trial. It is compulsory for each business unit to get answers to their operability concerns and cover each use case in detail for themselves.

As well as that, in order to reach a strong conclusion and rule out several providers, it would be helpful to speak to company references. Has anyone chosen the platform repeatedly over the years? Are there any success stories that the app company can provide at all? If the answer is no, then you might regard it as a major red flag.

Pricing

It is only rational to judge the overall applicability of a platform based on its pricing model. Simply put, software's total cost of ownership has to match your financial profit generation formula and potentially bring revenue to the business. Some might overlook investigating whether there would be some additional taxes, support, or hardware costs. You do not want to make that mistake and end up having a financial shortfall.

To conclude, it is advised to approach such optimization-related decisions with a practical mindset and try to narrow down solutions by creating a comparison table and possibly run demonstrations or trials. The value and insights in regards to conversion rates and customer behavior that web analytics offers are great. It has the potential to boost your business, but it must be based on your company's needs and future objectives. Take time to measure each solution side-by-side and study all the documentation and references if needed.

HOW TO READ ANALYTICS

As previously said, by using a web analytics tool, you can gain access to a wide range of data. But how do you transform and make sense of it? Most of the analytics solutions use a system known as ABC that stands for "Acquisition," "Behavior," and "Conversions." This means you can track the way your audience accesses the website, how they behave while on it, and what actions they take upon browsing. Even if you get reports full of raw numbers, it should not confuse you as it takes very little to interpret the data and determine its implications. Alternatively, you can apply the following ABC standard to analyze reports.

Acquisition

Various tools divide website visitors into different categories to simplify conducting standard analysis:

- **Organic search:** defines the number of people who have stumbled upon your website by typing relatable keywords in search engines. This indicator can reflect how good is your SEO strategy and whether your web content can attract enough interest.

- **Direct:** represents the number of visitors that have prior knowledge of your website and access it by typing directly your address or via a bookmark.

- **Paid searches:** the number of visitors acquired through running advertisement campaigns.

- **Social:** this stands for all the traffic directed from various social media channels.

- **Referrals:** defines a number of websites that have links and drive traffic to your site. This indicator might be helpful when establishing new partnerships or refining your communication strategy.

Behavior

Every website has a certain call to action incorporated in its display. Having access to analytics can let you monitor how people navigate and react to your site's message as well as how they decide whether to take the most desired action or not. The offered metrics have a lot of value if interpreted and put in the context correctly.

With web analytics you can see the following pieces of information about your web page:

- **Page views:** the number of times your website has been entered within a set period of time.

- **Unique page views:** represents the number of sessions during which that page was viewed one or more times.[3]

- **Average time on the page:** consequently, stands for the number of times a user spends on the website.

- **Bounce rate:** represents the percentage of people that chose to left your website right after accessing it.

- **Exit:** shows the percentage of visitors who click away to a different site from a particular page, after having visited any number of pages on your site.[4]

Conversions

Successful conversions happen when you reach main website goals. As previously stated, it is important that you clearly define those objectives in order to be able to track the company's overall efficiency. And if you are just starting to outline these goals, you may do so by grouping them into categories:

- Goals related to visitors reaching a certain section/content of your website

- Goals that state how much time you want visitors to spend on the website

- Goals revolving around a specific action—downloads, purchases, or subscriptions

---

[3] https://www.lens10.com.au/analytics-defintions-visits-visitors-pageviews-unique-pageviews/, Lens10

[4] https://canonicalized.com/bounce-rate-vs-exit-rate/, Canonicalized

CONCLUSION

Web Analytics is a broad domain that requires one to understand a huge range of concepts and terms such as unique visitors, page views, bounce rates, and conversion rates. Having correct analytics is really important for understanding what needs to be optimized in terms of your localization efforts as well. It allows you to establish an actual customer journey around your site and obtain more insights about design and content.

None of these metrics can tell the whole story of who uses your website and in what ways. But if you combine them and track them over the extended time period, you can get a much fuller picture that can help to introduce improvements to your website and SEO marketing. Another thing to keep in mind is the fact that website analytics is not a separate segment but a part of the overall digital communication field. It means that in order to have a neat communication flow, all the web analytic insights have to be paired with social media practices as well as SEO insights to produce more outcomes.

In the next chapter, we will cover other analytics and performance tools like file compression, minification, graphics optimization, databases, and hotlinks.

CHAPTER 6

Other Measures: File Compression, Minification, Graphics Optimization, Databases, Redirects, and Hotlinks

IN THIS CHAPTER

➤ Reviewing different measures of web optimization

➤ Studying various graphics optimization online tools

➤ Learning how to handle a database and prevent hotlinks

Website optimization is all about improving the performance of your website and your users' experience in order to meet their needs and eventually get more traffic. We have already discussed matters of website optimization in Chapter 2 for front end, content, and back end and Chapter 4 in regards to on-page optimization. So by now, it is hopefully clear that any optimization

DOI: 10.1201/9781003203735-6

that is not backed up with good data might only end up decreasing your conversion rate instead of increasing it. You need to know your target audience, learn their likes and dislikes, and give them the results that will maximize their satisfaction with the service. It would be correct to view website optimization as a scientific process where you would have to come up with a hypothesis for a change that might improve your conversion rate and then test it. And if you are right, you get more visibility and a well-tuned design, if you are wrong, you can always go back to original features.

And while optimization requires effort, the impact that it brings on your growth cannot be underestimated. It is also important to remember that search engine optimization (SEO) is a very dynamic and ever-changing field. And in order to stay updated about recent developments, you must be curious and always look for new sources of information. Popular industry websites like Search Engine Land https://searchengine-land.com/ or Social Media Today https://www.socialmediatoday.com/ are great platforms for the exchange of ideas as such.

Once you start applying optimization techniques without fear of failure or any hesitation, you sure to see positive results. This chapter aimed to help you with identifying measures like minification, graphics optimization, how to operate databases, and manage items like redirects and hotlinks.

IMAGE OPTIMIZATION

Most of the websites out there are comprised of images. Having images that you want to post optimized will reduce page load time as well as lessen the database space or mobile data volume. The process would also result in better search engine rankings and improved user experience due to increased speed and visibility.

Optimization methods differ based on your web structure and design preferences. The most popular image optimization practice is to "compress image size," converting them into lighter formats and reducing image details by removing some header information. Well-known and compression-friendly file formats that have static pixel color data are—png, .jpg, and .gif. Each format can be adjusted in terms of size and level of details, yet the trade-off of such compression might affect the overall image quality.

"Vector images" (e.g., .pdf, .ai, .eps) that use graphic parameters and coordinates are smaller than other formats and can be used to replace heavier images to decrease page load time, not having to give up on image quality. However, what needs to be mentioned is that vector images are mostly used for images that consist of geometrical graphics. It might be inappropriate and look unfit for those who want complex images or photographs on their websites.

"Image Caching" is a great solution that allows you to store image files in visitor's own browser cache enabling faster and safer access. It is considered the most effective optimization technique as it aims to reduce application requests rather than deal with image size and quality.

Proxy caching authorizes storing images on point of presence (PoP) servers, which helps to accelerate page rendering point. Caching is especially suitable for containing images that are shared among several pages as it lets visitors who have cleared their browser cache get faster access to images.

There are two ways to optimize images without having to compromise the quality:

- **Standalone tools:** These tools do not require any downloads or unnecessary installations. You just have to visit the website, upload the images you want optimized and later download them in a new format.

- **WordPress plugin:** In order to modify your images using these tools, you would need to install the plugin on your WordPress website and follow its basic settings for optimization.

Most users prefer standalone tools for their simplicity of operation, but they might be restricted in terms of image size or the number of images you can upload. Some of the popular applications you can try are:

- **TinyPNG:** TinyPNG uses fast optimization techniques that let you upload up to 20 images with a max of 5 MB each.
  Homepage—https://tinypng.com/

- **Compressor:** The tool supports JPEG, PNG, GIF, & SVG and allows you to reduce your image size up to 90% with the size limit of 10 MB.
  Homepage—https://compressor.io/

- **PunyPNG:** Is the most suitable tool for designers and developers. It can optimize for up to ten files at once with the limit of 150 KB per image.
  Homepage—http://www.punypng.online/

- **Compress Now:** This tool allows you to choose the level of compression, the percentage of weight loss that you need.
  Homepage—https://compressnow.com/

- **Kraken:** The tool is defined as: "a robust, ultra-fast image optimizer and compressor with best-in-class algorithms. We'll save you bandwidth and storage space and will dramatically improve your website's load times."
  Homepage—https://kraken.io/

- **Imagify:** Imagify gives you three compression options to choose from—normal, aggressive, and ultra. It also stores your files for 24 hours so you can access them again if needed.
  Homepage—https://imagify.io/
  Now let's take a look at the following WordPress plugin optimization tools.

- **EWWW Image Optimizer:** EWWW Image Optimizer is usually preferred for running a photography-related website. It allows you to set your own specific standards for image reduction in size and quality.
  Homepage—https://wordpress.org/plugins/ewww-image-optimizer/

- **WP Smush:** WP Smush lets you compress not only single images but also those coming as a set; yet has a limit of a maximum of 50 free compressions.
  Homepage—https://wordpress.org/plugins/wp-smushit/

- **Optimus:** Optimus has the limitation of 100 KB per file but allows you to reduce the image size by up to 70%.
  Homepage—https://wordpress.org/plugins/optimus/

- **Short Pixel:** It compresses GIF, PNG, JPG, & PDF files and is compatible with WooCommerce open-source e-commerce plugin.
  Homepage—https://shortpixel.com/

JAVASCRIPT PERFORMANCE

JavaScript (JS) is an essential feature of practically every web-based or mobile app software. And while JS has the capacity to make your website more appealing, the accidental poor script can just as well ruin your users' experience. We shall further explore the causes of JS performance deficiencies and provide a list of the best tools for optimizing your JS.

Any performance improvements start with identifying the cause of regular complications you might be facing. There are few common setbacks when integrating JS into your web structure:

1. **Too many interactions with the host:** Too many interactions with the host object or the user's browser contribute to blocking or slowing the Document Object Model (DOM) programming interface.

2. **Too many dependencies:** If you have plenty of JS dependencies, then the speed and quality of your application's performance will be affected, forcing your users to wait longer for page items to load.

3. **Messy code:** A lack of systematic structuring of your JS code can result in the inadequate allocation of tools and resources.

4. **Poor event handling:** If you do not use event handlers properly or do not improve them from time to time, they can start operating repeatedly without your knowledge causing more confusion and potential errors.

Fortunately, there is a number of helpful practices for improving JS performance. Some of them include

1. **Use HTTP/2:** Switching to the latest version of the HTTP protocol will not only improve your JS performance but is likely to speed up your site in general. It also has great factors like multiplexing that

allow multiple requests and responses to be sent simultaneously, therefore accelerating the loading speed.

2. **Clean your HTML:** You can start trimming your HTML by simply getting rid of unnecessary <div> and <span> tags. And shall you achieve sizing down your code, you will definitely see how it doubles your DOM speed.

3. **Minify your code:** Cleaning your HTML or minifying your JS code can get you faster, more accurate performance. We are going to cover minification in detail further in this chapter.

4. **Compress your files:** Using the same logic as compressing your images, smaller size JS will enable faster downloads and improved page performance. You can use free, online compression tools such as Gzip https://www.gnu.org/software/gzip/ or Brotli https://github.com/google/brotli to reduce the size of your JS.

5. **Limit library dependencies:** Using library dependencies affects your web loading time. You can try using browser-based instead of relying on external store platforms.

6. **Use code splitting:** First Contentful Paint (FCP) measures how long it takes the browser to render the first piece of content after a user navigates to your page. One of the best ways to achieve a higher FCP score is to use code splitting. Code splitting is a technique where you send only the necessary modules to the user in the beginning. This would greatly impact the FCP score by reducing the size of the volume transmitted initially.

7. **Use async and defer:** By default, the browser must wait until the script downloads, execute it, and then it processes the rest of the page. This can lead to your bulky script blocking the loading of your webpage. In order to escape this, JS provides us with two techniques known as async and defer. You have to simply add these attributes to the <script> tags.

   Async enables the browser to load your script without affecting the rendering. In other words, the page doesn't wait for async scripts, the contents are processed and displayed.

Defer tells the browser to load the script after your rendering is complete. If you specify both, async takes precedence on modern browsers, while older browsers that support defer but not async will fall back to defer. These two attributes can greatly help you reduce your page loading time.[1]

8. **Cache, cache, cache:** One of the things that actually help you to speed up your website is caching. Make sure that both your browser cache as well as intermediary content delivery network (CDN) run correctly.

If you need support in code compression or even code minification there are great, accessible tools designed to help developers optimize their JS performance:

1. **Google Closure Compiler:** It is "a tool for making JS download and runs faster. Instead of compiling from a source language to machine code, it compiles from JS to better JS. It parses your JS, analyzes it, removes dead code, and rewrites and minimizes what's left. It also checks syntax, variable references, and types, and warns about common JS pitfalls."[2]

   Homepage—https://developers.google.com/closure/compiler

2. **Packer:** Packer is a highly effective online Java compressing and decompressing tool.

   Homepage—https://docs.oracle.com/javase/8/docs/api/java/util/jar/Pack200.Packer.html

3. **Dojo ShrinkSafe:** ShrinkSafe is a "standalone Java-based JS compressor which utilizes Rhino to parse code and safely shorten the results. ShrinkSafe renames local references to short names prefixed with an underscore. This saves bytes on the wire and also provides some obfuscation of the code. It also eliminates whitespace and comments when generating the new code."[3]

---

[1] https://blog.bitsrc.io/14-javascript-code-optimization-tips-for-front-end-developers-a44763d3a-0da, Blog Bitsrc

[2] https://developers.google.com/closure/compiler, Google

[3] https://dojotoolkit.org/reference-guide/1.9/util/shrinksafe/index.html, Dojo Toolkit

Homepage—https://dojotoolkit.org/reference-guide/1.9/util/shrinksafe/index.html

4. **YUI Compressor:** YUI Compressor is a tool created by Yahoo! that also compresses CSS files and has a very high compression ratio. Homepage—https://yui.github.io/yuicompressor/

It may be challenging to grasp all of these recommendations at first, but the sooner you get the habit of implementing these practices, the less likely you will ever deal with JS errors later on. And shall you choose to use any of the tips individually, it will result in a slight performance uplift. But if you can implement them together, you are more likely to notice significant improvements in your web operations.

VIDEO OPTIMIZATION

Adding multimedia of all sorts on your website is a great way to engage with your audience. Video content in particular is in high demand. First of all, videos provide a great audio-visual experience to your users: they could be short in time but deliver the key information in a flashy and entertaining manner. Effectively produced videos also capable of giving your site and brand their own unique identity. You can use this opportunity to tell customers more about your team and a personal story behind your company. Aesthetic videos that focus more on your products or services could also help to show off your style to the web. Video content is also considered to be perfect for social media engagement. Once your videos are ready and featured on your social media accounts, you are going to have a lot of attention as well as redirected web traffic.

Additionally, videos also affect your SEO. Visitors are more likely to browse pages that have various multimedia presentations on them, not just plain text. That is why search engines rank websites that match user's preferences in design and multimedia much higher than those offering script-based content.

However, in order to deliver that heavy video content that you know your visitors will like, it has to be properly optimized first. Standard high-quality video content optimization is a two-step process. First, you need to take care of the video file, so it does not affect your site's page speed. And then you need to optimize the web page to be able to integrate the video content properly.

The first step requires the following actions:

1. **Use HTML5 supported formats:** In the early days, you had to deal with browser plugins in order to install video players like Windows Media Player or RealPlayer. That with time progressed to Quicktime or Flash, which lacked speed and data protection policies. We now use the universal programming language used in creating web pages the HTML5 that enables multimedia enhancements like HTML5 video encoding. With that, you can encode videos through simple procedures, and the quality of media and geolocation of devices have also advanced.

   HTML5 video encoder works to allow users to upload a video in its original format and then automatically produces a complete HTML5 encoded file without the need for anyone to manually type the code. Moreover, HTML5 video converter also allows to upload files in multiple quantities and will proceed to convert them with an effortless press of a button.

   Some online HTML5 online video encoders include

   - Converter Point—https://converterpoint.com/

   - Miro Video Converter—http://www.mirovideoconverter.com/

   - Handbrake—https://handbrake.fr/

   - Freemake Video Converter—https://www.freemake.com/

   - Online-convert—https://www.online-convert.com/

2. **Compress video data with tools:** No matter what your website is about, you want to add a certain appeal to it by adding high-quality high-definition (HD) or 4K videos. However, in most cases, in order to accommodate a file like that, you would need to compress it a little at first. Fortunately, online compression tools and affordable software can shrink the size of files without compromising the quality.

   For compressing a video for websites without software, you can use Media.io. that edits both online video and audio—https://www.media.io/. Another option of compressing your video would be through video editing software like Windows Movie Maker, Adobe Premiere, iMovie, Adobe Media Encoder.

3. **Use a CDN:** We have previously defined and covered the necessity of a CDN responsible for the fast delivery of content, even if bandwidth is limited, or if there's a sudden spike in website traffic. Because a CDN finds the closest server to the requesting end-user, it removes the distance from the content delivery equation. It also reduces the number of hops that data makes and therefore considered essential users across the world, most of them being online business corporations.

For this reason, websites that rely on heavy video content a lot have to get theirs own CDN provider. Ultimately, it is not only going to improve the speed of multimedia delivery but affect your website's overall performance for the better.

4. **Mobile-first optimization:** People browsing the web through their mobile devices are on the verge of completely overtaking the number of people accessing the web from their desktop computers. Google's mobile-first indexing has made the even bigger significance of this phenomenon. It is becoming more apparent that failing to timely optimize the content of your website for mobile users might put your business at a huge risk.

Thus, it is important to display videos on the website in the size and width ratio that mobile devices will be able to handle. Developers also recommend creating a different background for mobile-friendly, fast operations as no one wants their visitors running out of mobile data while viewing your website.

5. **Build calls to action:** Before you release your next video, make sure you have included specific calls to action.

Most call to action buttons (CTAs) typically include series of YouTube videos pointing to different resources, direct prompt from the video's host, or a short link at the end of the video directing viewers to a landing page. Your CTAs should not only be direct, but they should also include different ways that your audience might want to reach out to you.

6. **Set up a lead-capture method:** Another feature that's moved directly from the written content world into the video content world is the email gate and lead-capture form. Once you are ready to upload your video, it is important to consider including an email gate because

these will go wherever your content is shared on social networks, providing a simple way to find out which leads are interested enough to willingly give you their personal information.

7. **Add key information for SEO:** SEO can be a tricky beast, in part due to the frequent algorithm updates from Google; but making your videos more search-friendly is easy if you focus on these three things:

- **Keywords and descriptions:** make sure you do some initial research on the words you would like to rank for within your industry and use those words in a clear phrase format for your video's title and in your meta descriptions.

- **Transcripts:** to help with video SEO, try transcribing your video (or use a service) and turn the video's accompanying text into a blog post. This way the blog post featuring your video will alert search engines about the context, and Google will qualify your relevant content.

- **Multi-platform promotion:** after you have embedded your video on your site, put it up on YouTube and other distribution outlets with a different title.[4]

To conclude, what needs to be emphasized once again is the fact that every website has a different target audience. And it is an audience that has its wants and needs. If the video content you are posting adds value to your customer's web experience, then optimizing it will pay off a huge deal. But if your videos are too slow and do not enhance your best features, then going through the above-mentioned recommendations is not really worth it.

When it comes to optimizing your media files, you must track the effect that it brings to your website conversion rates. Any multimedia content has to be used strategically and regularly reevaluated to ensure that every item on the display serves a purpose. Do not hesitate to test new techniques and try something new.

---

[4] https://contentmarketinginstitute.com/2014/02/optimize-video-content-simple-process/, Content marketing institute

MINIFICATION

Wikipedia defines minification as "the process of removing all unnecessary characters from the source code of interpreted programming languages or markup languages without changing its functionality. These unnecessary characters usually include white space characters, newline characters, comments, and sometimes block delimiters, which are used to add readability to the code but are not required for it to execute."[5]

It is common that developers mostly use spacing and comments when creating HTML, CSS, and JS files. And while it is useful when working with coding, it later becomes a negative setback when it comes to serving websites because not having a well-structured code generates additional network traffic.

Minification is also one of the main methods used to reduce page load times and improve site speed and accessibility. To minify JS, CSS, and HTML files, comments and extra spaces need to be erased; and variable names resized so as to minimize code and file size. With that, minimized file version will still provide the same functionality.

Nowadays, when most of the major JS library developers provide minified versions of their files for further implementation, minification has become a standard application for page optimization. However, most avoid that practice as it could be heavy and unmanageable even when implemented with automated tools as you will probably need to keep separate file versions anyway. Employing a CDN can take the burden of minifying your own files as it has the option of automated minification. Even after optimizing the code, CDN will still keep your original, uncompressed files on the main server, while storing minified version on its caching servers.

If you are looking to minify your HTML, CSS, and JS files hassle-free you should try:

- **Closure Compiler (JS only):** It allows you to minify your JS along with other helpful optimizations. You can pull in your JS using by entering the URL of the js file location and then choose how you want the code to be optimized and formatted. https://closure-compiler.appspot.com/home

---

[5] https://en.wikipedia.org/wiki/Minification, Wikipedia

- **cssminifier.com and javascript-minifier.com (CSS and JS):** These two minifiers by Andrew Chilton are very easy to use. You can simply paste in your code and then click the Minify button to output the minified code. https://cssminifier.com/ and https://javascript-minifier.com/

- **csscompressor.net (CSS only):** Another quick, easy, and free to use online CSS compressor. https://csscompressor.net/

- **jscompress.com (JS only):** This JS compression tool allows you to compress JS code via Copy and Paste but you can also upload multiple JS files at once. This is great for combining JS files into one file for better page load speed. https://jscompress.com/

- **refresh-sf.com (HTML, CSS, and JS):** It also includes all the compressor options for JS, CSS, and HTML code types. https://refresh-sf.com/

- **htmlcompressor.com (HTML, CSS, and JS):** This online compressor/minifier tool also supports HTML, CSS, and JS code types. It even supports different combination of code types like CSS + PHP and JS + PHP. It allows you to check the compressed code for errors as well. https://htmlcompressor.com/compressor/

- **minifycode.com (HTML, CSS, and JS):** This minifies offers solutions for JS, CSS, and HTML with a simple and clean UI that minifies your code with a single click of a button. http://minifycode.com/

DATABASES

We, as a human collective, produce and accumulate an enormous amount of data per second. And now we find ourselves dependent on cloud databases to hold that collection of facts and figures in an organized order. There are two different types of digital database management: relational and web databases. A "relational database" is specialized in storing data in groups or tables and has the ability to connect and relate several files together. It does so by applying an indexing method in which data is added with specific keys to it that help to locate relatable information fields stored in the same database and regain it quickly.

A "web database" is basically web storage that can be accessed from a local network or the Internet. Unlike regular databases that are attached to

computed memory, web databases hosted on websites, and are considered to be software as service (SaaS) products. Content management systems frequently use web databases to store site records such as multimedia, posts, profile information, and comments. This database is accessible for anyone, even those who do not want to get involved with actual programming as it offers easy updates without the need to edit the HTML code at all. If designed properly, web databases can expand and host millions of entries on a daily basis. It is also a great solution to reduce the web's memory utilization and boost the speed of operations.

Some advantages of using a web database include

- Web databases can be multifunctional. You can use it for both professional and personal purposes. Common ways e-commerce businesses use their web databases are for survey forms, customers' profiles and data, product documentation, and staff lists. Personal web databases are a great way to store your memorabilia, documents, or email addresses.

- A web database is the fastest way to retrieve your data. Unlike a spreadsheet or any file system, the database is designed to manage large volumes of data. It has a very complex structure that it optimized to the maximum to deliver fast search results.

- Information that you store in a web database can be accessed from almost any device. Having digital storage means your files are not tied to just one computer. With a web database, you can get a hold of your data from any place in the world or any device as long as you are granted access.

- Web databases have a great technical support team that is at your disposal shall you face any issues or have concerns regarding data security. They can offer services such as data-oriented and user-oriented security, administration software, portability, and data recovery support. There is almost no need for your IT department to get engaged with database matters personally.

- Web databases are very convenient to update, download or reload any information. The process is navigated in a very simple manner.

You should definitely consider opting for a web database service if the data that you are utilizing has one of the following characteristics:

- There is more than one kind of data subject. For example, there might be information about customers, accounting details, inventory, and other data.

- There is a simply large amount of data.

- There are multiple users who need to access the data at the same time.

- There are relationships between the stored data objects.

- Consolidated reports must be produced from stored, related information; that is, the data must be obtained and analyzed to produce results.

- Adding, erasing, and updating data is a time-consuming, complex process.

There are various ways by which databases can be tuned:

1. **Proper indexing:** Index is basically a data structure that enables the data retrieval process overall. Indexing that creates separate data columns without overlapping each other ensures quicker access to the database. Excessive indexing or no indexing at all are both trigger ineffective database usage.

2. **Using or avoiding temporary tables according to requirement:** If any code can be well written in a simplified way, there is absolutely no need to make it complex with temporary tables. Only in case, a data has a specific procedure to be set up which requires multiple queries, the use of temporary tables in such cases are, in fact, recommended.

3. **Avoid coding loops:** Avoiding coding loops could be very useful in order to avoid slowing down of the whole sequence. This can be achieved by using the unique UPDATE or INSERT commands with individual rows, and by ensuring that the command WHERE does not update the stored data in case it finds a matching preexisting data.

Depending on who uses the database and for what purpose, it can be either a static or dynamic solution. For any entity that has even moderate structure and data complexity like e-commerce sites, social networks, or blogs, it is important to switch to a database management system to preserve and operate files in an error-free manner. At the moment, there are three commonly used database management systems out there:

MySQL

MySQL is one of the most popular database management systems for the Internet. It is an open-source solution that is used not only as a relational database but also for logging applications and e-commerce. The MySQL name stands for "My," the name of co-founder Michael Widenius's daughter, and "SQL"—Structured Query Language.

The database has been integrated with a set of open-source software applications known as LAMP stack, which stands for Linux server operating system, Apache web server, MySQL as the database, and PHP programming language. Many of the world's largest corporations like Facebook, Adobe, and even Google choose MySQL as their database solution due to its great package deals and ability to store and process high-volume information.

Major features as available in MySQL 5.6:

- Cross-platform support

- Triggers

- Cursors

- Updatable views

- Online Data Definition Language (DDL)

- Performance Schema that collects and aggregates statistics about server execution and query performance for monitoring purposes

- A set of SQL Mode options to control runtime behavior, including a strict mode to better adhere to SQL standards

- Query caching

- Built-in replication support

- Full-text indexing and searching

- Embedded database library

- Partitioned tables with pruning of partitions in optimizer

- Multiple storage engines, allowing one to choose the one that is most effective for each table in the application

- Commit grouping, gathering multiple transactions from multiple connections together to increase the number of commits per second

Homepage—https://www.mysql.com/

PostgreSQL

PostgreSQL is a "powerful, open-source object-relational database system that uses and extends the SQL language combined with many features that safely store and scale the most complicated data workloads."[6] It has great features like multi-factor authentication, customizable storage interface for tables, advanced Indexing, and multi-version concurrency control. Moreover, PostgreSQL functions on all major operating systems and currently works on its own PostGIS geospatial database extender.

---

[6] https://www.postgresql.org/, PostgreSQL

A wide variety of native data types are supported, including:

- Arbitrary-precision numerics

- Character (text, varchar, char)

- Binary

- Date/time (timestamp/time with/without time zone, date, interval)

- Money

- Text search type

- Composite

- Arrays (variable length and can be of any data type, including text and composite types) up to 1 GB in total storage size

- Geometric primitives

- XML supporting XPath queries

- Universally unique identifier (UUID)

Homepage—https://www.postgresql.org/

MongoDB

MongoDB is probably the most popular and open-source non-relational data management system. It defines itself as a "general-purpose, document-based, distributed database built for modern application developers and for the cloud era."[7]

Instead of following relational database design, MongoDB has introduced a different method of storing data as a collection of files, using a simpler layout that can be easily be computed using an automatic sharing feature. This modern approach to store data is said to be more natural for programmers and is very appreciated by JS developers because it has replaced with JSON standard format documents. Thus, it enables complex data to be added to the database in any structure and MongoDB adapts it automatically before transmitting.

Homepage—https://www.mongodb.com/

---

[7] https://www.mongodb.com/, MongoDB

The above-mentioned list of commonly used database management systems does not intend to say that no other solution should be used. Developers usually have their own preferences regarding what might work better for a specific project and could help solve the problem better. Therefore, for better shopping for a database, you must think as to what issue or cause you to want to solve with this tool. It is also encouraged to try and test those systems to see their features and have the chance to compare data stored routes. A good test drive and analysis like that will make your long-term investment safer.

REDIRECTS

The HTTP redirect code is just a way to forward search engines and direct visitors from one URL to another. Wikipedia describes redirects as a "World Wide Web technique for making a web page available under more than one URL address. When a web browser attempts to open a URL that has been redirected, a page with a different URL is opened; similarly, domain redirection or domain forwarding is when all pages in a URL domain are redirected to a different domain."[8] Redirects are usually used when a website is relocating to a new domain, temporarily unavailable due to server maintenance, or merging content.

And even though it is a great practice to be able to forward your website visitors from old to updated content on new pages, it is still recommended

---

[8] https://en.wikipedia.org/wiki/URL_redirection, last edited on December 4, 2021

to avoid using redirects if possible. But if that is the only option, you need to make sure to follow these useful tips:

- Redirects should be structured using the right protocol (HTTP or HTTPS), the domain name (www or non-www), and path notation (with or without trailing slash).

- One redirect should not forward your visitors to another redirect.

- Use a 302 redirect for irrelevant content.

- Use a 301 redirect for content that is permanently removed.

Redirects give chance to maintain a good standard of user experience. Thus, you do not want to confuse your visitors with a 404-page. You want to think about their journey in advance and smoothly redirect them to the available webpage. Just like people, search engines cannot understand what could happen to your content, whether it was temporarily moved or permanently erased. Implementing redirects should be viewed as sending the right warning message to avoid disorientation and secure your SEO ranking.

When to Use Redirects

As previously stated, there could be different reasons behind why you would have to apply redirects. Some of those causes are

- **Deleting pages with SEO value:** In case you are running a catalog on your website and some products will no longer be available, you can use 301 redirects to forward your audience to another URL with the most relevant alternative item. It could be a great solution to enhance customer service and keep visitors on your page.

- **Switching domain names:** Moving to a new domain name is a common practice that may happen to anyone, but when doing so, you need to make sure you redirect both visitors and search engines to a new domain ID.

- **Merging websites:** Another situation where you could use a website redirect is when merging websites. It could happen if you decide to merge a few of your websites into one and rebrand it. You would need to then redirect your traffic to a new page using 301 redirects.

The reason why you would want to avoid redirects is that it can result in space waste and loss of link worthiness. It is therefore a key priority to keep your internal links on point whenever optimizing the web.

HOTLINKS

Hotlinking is the act of linking to a file from a third-party website, instead of properly downloading it on your own server with added citation. The most commonly linked items are photos or images as well as audio files and various animations. Hotlinking is customarily considered bad etiquette sometimes people go further and call it theft.

Many photos you see around the web have licensing regulations attached to them, saying that commercial usage or reposting are not permitted. People who still choose to link to those pages very likely do not have the authorization to break the above restrictions. This also means that they do not have to pay for the license or make a reference to the original source. Hotlinking can also be a burden and loss of resources for the initial server. If someone links to your images, the number of your search queries could go up and cause delays and even suspension of your hosting account. In order to prevent such parasitism, you need to secure a high-performance host that can prevent such instances, and acquire effective hotlink protection software.

Nevertheless, there are cases when hotlinking occurred unintentionally, and the hosting serves were not even aware of the incident. It could happen while someone was copy-pasting URLs and files to create a new code or because they did not know how to properly link to multimedia. However, this could not be a reason for such exploitation of resources. And if you take on writing content for the web, you should at least know its best practices and the worst practices.

How to Prevent Hotlinking

Below is a number of preventive measures you can implement to stop users from hotlinking your data. Not practices such measures could leave your website vulnerable to external (even if unintentional) harm:

1. **Use a CDN with hotlink protection:** The simplest and safest way to prevent hotlinking is to opt for a CDN that has a built-in hotlinking protection feature. Once switched on, this feature will block websites from linking to your image resources without permission. They

would still be able to view and download images from your website, however. And in case some would really want to use your files, they would be forced to properly host, and cite images from your website. Cloudflare is a good example of a platform that offers that kind of hotlink protection.

2. **Disable right-click functionality:** In order to hotlink an image, users would have to first right-click the image, copy the image address, and set the URL onto their website. And to prevent such a procedure, the easiest you can do is to disable the right-click functionality altogether. On the other hand, it is surely not going to stop those who might be determined to do it as there are many ways to go around this obstacle. But it may work for users who hotlink accidentally, without bad intentions. In that case, it might be a helpful recommendation to follow.

3. **Add a watermark to your images:** Another solution that is not likely to stop every user from hotlinking but may help to send the warning message is to add a watermark to your images and other multimedia. You can do so by simply attaching a watermark that can contain your name, copyright trademark, and maybe a logo if you have one. In case you want to emphasize your ownership in a stronger manner, you may also add information about the penalty that one might face. If you need online help with your watermarks you can use a free tool like Watermarkly https://watermarkly.com/, or Adobe Lightroom Classic https://www.adobe.com/products/photoshop-lightroom-classic.html.

4. **Rename hotlinked files:** Another quick fix one may apply is renaming files. If you notice that your images might be hotlinked by other websites, changing file names will cause those hotlinks to break and redirect users to 404 errors.

5. **Issue a takedown request:** A more official approach to deal with hotlinked resources on your site is to file a takedown request. A takedown request or so-called Digital Millennium Copyright Act (DMCA) notice has to be forwarded to a company, web host, or search engine stating that they are or linking to your data that is protected by a copyright notion.

Once the other party receives the notice as a hotlinker, they are expected to immediately remove the material from their website or to add property citations to downloaded files. Otherwise, their Internet service provider has the right to consequently fine, block, or suspend their website. Knowing that most companies usually remove the hotlinked content rather promptly.

6. **Block IP addresses:** If you are making use of your web analytics, you would be able to notice if a large portion of your traffic would be coming from only a few sites. This is usually one of the evident signs that they could be hotlinking images or other data on your site. If you suspect that to be true but do not want to go through the hassle of filing a complaint, you can start to try renaming the files or blocking their IP address. Blocking IP addresses could be a great final act if the other source has been ignoring your polite warning messages for a while now.

Taking mentioned steps to prevent hotlinking is important to protect your web files from intentional or accidental abuse as well as ensure you get well-deserved credit and conversions for media files that you produce. Unfortunately, whether you have hotlink protection or not; or whether your website has been hotlinked does not influence the website ranking at all. It means that these issues do not get much focus and protection against such instances is not properly incentivized yet. In the final chapter, we are going to talk about in-depth server-side optimization and see how it can prevent cases of misuse and violation that we have covered.

CHAPTER 7

Server-Side Optimization

IN THIS CHAPTER

> Understanding a web server and different types of web servers

> Studying operating systems for servers

> Learning server-side optimization methods

A web server is a software and hardware that utilizes Hypertext Transfer Protocol (HTTP) and other protocols to respond to user requests over the World Wide Web. Its main function is to store and deliver the content for a website like text, multimedia visuals, and other application data. Every computer that hosts websites has server software installed. And while web server hardware is connected to the Internet, which allows data to be interchanges with other devices, the web server software drives to ensure user access to hosted files. The web server proceedings are based on the client/server operating model. Besides HTTP, web servers also support Simple Mail Transfer Protocol (SMTP) and File Transfer Protocol (FTP), used for email transfer and storage.

Web servers are usually used to publish webpages, download requests for FTP files, and email interaction. You can connect to web server software through the domain names of websites. For instance, when Google Chrome web browser requests a file that is on a web server, it will request the file by HTTP. And once the web server receives the request, the HTTP

DOI: 10.1201/9781003203735-7

server will accept it and send the content back to the browser through HTTP. More specifically, that process follows an array of intertwined steps. First, a user specifies an URL address, the web browser then obtains the IP address of the domain name through the URL domain name system (DNS) or by searching in its cache. The browser then requests the specific file from the web server by an HTTP as previously mentioned. If the web server responds, the content will be sent through HTTP. And in case the requested page does not exist, the web server will respond with an error message.

Basic server-side scripting runs on the server machine and has a wide-range feature set, which includes databases. It typically uses Hypertext Preprocessor (PHP) or Active Server Pages (ASP), scripting languages. Additionally, there are two types of content-driven by a web server—static and dynamic. Static refers to the content being demonstrated and sent to the browser unchanged as is. While dynamic content consists of other software such as an application server and database. It is therefore used to update any hosted files before they are sent to a browser.

In this chapter we will cover the following three top web server software on the market right now:

- **Nginx:** A popular open-source web server that has event-driven architecture and is known for its utilization and scalability.

- **Litespeed:** A free web server is seen as one of the fastest because of lesser central processing unit (CPU) power consumption.

- **Apache HTTP Server:** An open-source web server is used for Windows, macOS X, Unix, Linux, Solaris, and other operating systems (OS).

Each one of those has different abilities to handle server-side programming, site-building tools, and security characteristics. They also vary in configurations and set of default values. When considering which high-performance webserver to opt for, you should examine how well it works with the OS and other servers.

NGINX

NGiИX

Nginx is an open-source web server that was released in October 2004. And since its initial success has also been upgraded to serve as a reverse proxy, HTTP cache, and load balancer. Igor Sysoev created Nginx as an answer to the C10k problem regarding the performance issue of handling 10,000 concurrent connections. And now has proven so effective that many high-profile companies like Google, Adobe, Xerox, Autodesk, Microsoft, IBM, Facebook, Target, Twitter, and Apple choose it over other server solutions.

Nginx default configuration file is nginx.conf, usually found in /usr/local/nginx/conf, /etc/nginx, or /usr/local/etc/nginx.

HTTP proxy and Web server features[1]

- Ability to handle more than 10,000 simultaneous connections with a low memory footprint (~2.5 MB per 10k inactive HTTP keep-alive connections)

- Handling of static files, index files, and auto-indexing

- Reverse proxy with caching

- Load balancing with in-band health checks

- TLS/SSL with SNI and OCSP stapling support, via OpenSSL

- FastCGI, SCGI, uWSGI support with caching

- gRPC support since March 2018, version 1.13.10

---

[1] https://en.wikipedia.org/wiki/Nginx, last edited on December 9, 2021

- Name- and IP address-based virtual servers

- IPv6-compatible

- HTTP/1.1 Upgrade (101 Switching Protocols), HTTP/2 protocol support

- URL rewriting and redirection

Nginx in most instances like benchmark testing or running static content claims to outperform other popular web servers because of its roots in performance optimization under scale. With Nginx, you can tune almost any setting, but the rule to follow should be to change one setting at a time, and bring it back to the default value if the change does not improve functioning.

Many like to compare Nginx to Apache, but for many application types, Nginx and Apache complement each other well. A very common starting pattern is to deploy Nginx Open Source as a proxy in front of an Apache-based web application. Nginx performs the HTTP-related heavy lifting—serving static files, caching content, and offloading slow HTTP connections—so that the Apache server can run the application code in a safe and secure environment.[2]

Here we shall cover how you can tune your Nginx configuration to bring more benefit under normal workloads.

The Backlog Queue

The Backlog settings correspond to connections and how they are queued. It is worth checking on these settings if you have a high number of incoming connections and getting irregular or uneven performance levels.

- **net.core.somaxconn:** stands for the maximum number of connections that can be queued to be received by Nginx. The general default number is often very low, and if you are experiencing disconnections during heavy traffic hours, or if the kernel log indicates that the value is too small, this number can be increased.

---

[2] https://www.nginx.com/blog/nginx-vs-apache-our-view/, Nginx

- **net.core.netdev_max_backlog:** shows the rate at which the network card buffers packets before being transferred to the server CPU. Increasing this figure can improve performance on machines with a high amount of bandwidth.

File Descriptors

File descriptors are OS resources used to stand for server connections and open files. Nginx can use only up to two file descriptors per connection. Namely, if Nginx is proxying, one file descriptor would be used for the client connection and another for the connection to the proxied server. And if you are serving a large number of connections, you might want to adjust the following settings:

- **sys.fs.file-max:** stands for the server-wide limit for file descriptors

- **nofile:** used to tune user file descriptor limit, could be set in the/etc/security/limits.conf file

When Nginx is running as a proxy and each descriptor-established connection is temporary, you need to check:

net.ipv4.ip_local_port_range—represents the start and end of the range of port values. If you observe that you are running out of ports, you may need to increase this range.

Worker Processes

Unlike file descriptors, worker processes can operate in multiple quantities, each processing a large number of simultaneous connections at the same time. The quality and speed of these processes could be controlled with the following directives:

- **worker_processes:** consequently, shows the number of Nginx worker processes. In cases when you are running one worker process per CPU core, no adjustments are necessary. However, if you are experiencing a heavier workload, you might want to increase this number.

- **worker_connections:** stands for the maximum number of simultaneous connections that one worker process can handle. The default number is 512, but if your traffic runs higher than usual, Nginx has enough resources to support a larger number.

Keepalive Connections

Keepalive connections support Nginx clients and upstream server links. Users generally open a number of simultaneous connections and conduct keepalive transactions across all of them. These connections could be held available until the user shuts the browser or the browser would time them out. Modern web browsers can handle around six–eight keepalive connections at the same time. Tuning the following connections can have a major impact on the overall performance of the CPU:

- **keepalive_requests:** the number of requests the user makes over a single keepalive connection. The default value is 100, but you can test whether a higher number can be hosted with the load-generation tool.

- **keepalive_timeout:** stands for the amount of time that keepalive connection can remain open.

- **keepalive:** the number of keepalive connections to an upstream server that stay open for each worker process. This could be tuned without the notion of the default value.

Access Logging

Logging every request consumes Input/Output scheduling and slows down the processing unit. In order to reduce that impact, it is recommended to enable access-log buffering. That way, instead of performing a separate operation with each log entry, Nginx buffers a series of entries and hosts them together in a single operation.

To enable access-log buffering, add the buffer size parameter to the access_log directive; and if you want Nginx to write the buffer after a specified amount of time, insert the flush=time parameter. On the other hand, to disable access logging completely, include the off parameter to the access_log directive.

Sendfile

Sendfile system is mostly used to speed up data transfers as it copies data from one file descriptor to another. It is not, however, a part of the regular Nginx processing chain, and to enable Nginx to use it, you should include the send file directive in the HTTP context or location context. With Sendfile on, Nginx can then write cached or on-disk content without any context switching to userspace, making it less jammed and more accelerated.

Monitor Your Nginx

If you want to avoid any potential worry or hassle, you need to regularly monitor your Nginx server. It is important to get first-hand updates on performance optics and find out potential errors before your users experience them. You can use one of the most effective monitoring solutions for Nginx such as The Nginx monitoring extension for AppDynamics (https://docs.datadoghq.com/integrations/nginx/?tab=host), Dynatrace application (https://www.dynatrace.com/technologies/nginx-monitoring/) or the Nginx plug-in for Pingdom Server (https://server-monitor.pingdom.com/plugin_urls/72-nginx-monitoring).

The above-mentioned directives guarantee to bring impact to Nginx performance. We have chosen to mention those that are safe to adjust on your own. Shall you change the server's basic configurations, we recommend that you do it with help and guidance from the Nginx support team.

LITESPEED

LiteSpeed is another world-leading high-performance web server. It was founded in early 2002 by a team of engineers led by George Wang. On July 1, 2003, LiteSpeed Web Server was officially released as a full-featured web server. was founded in early 2002 by a team of engineers led by George Wang. On July 1, 2003, LiteSpeed Web Server was officially released as a full-featured web server. And in December 2020 was the fourth most popular web server, estimated to be used by 8.1% of websites.[3]

It has a wide range of features and an uncomplicated web administration console. Litespeed is appreciated for its capacity to replace an Apache

---

[3] https://w3techs.com/technologies/overview/web_server, W3Techs

server without changing any OS details. Therefore, it can quickly solve any existing web hosting platform and help maintain an effective web hosting infrastructure. The web server is available in three editions:

- **OpenLiteSpeed edition:** it is an open-source, free edition made for personal use. It can operate medium to high-traffic websites but is not compatible with any hosting control panel.

- **Standard edition:** this edition, on the other hand, is compatible with hosting control panels such as WHM/cPanel & DirectAdmin. It is used for small, low-traffic websites and is also free and can be used for personal or commercial purposes.

- **Enterprise edition:** this edition has the highest level of stability as it is widely used by leading web hosting organizations. It runs large, high-traffic websites and is compatible with multiple hosting control panels.

LiteSpeed is considered to be faster than Apache while serving PHP content. It is also preferred for serving WordPress, Joomla, and Drupal-based websites and has other beneficial features such as:

Apache Compatibility

LiteSpeed is compatible with the commonly used Apache web server and can even load the Apache configuration files, sometimes performing as a direct replacement for Apache. It is also greatly compatible with the leading hosting control panels like DirectAdmin, cPanel, and Plesk.

Event-Driven Architecture

Due to its event-driven architecture, LiteSpeed can handle sudden spikes in traffic as well as manage against denial-of-service (DDOS) attacks. Additionally, it can serve thousands of clients simultaneously using a minimum of server memory and increase PHP performance to serve static websites faster than Apache.

Security

Because LiteSpeed has customizable features like bandwidth throttling and per-IP connections, it can block IPs that request too much bandwidth or make too many simultaneous connections are blocked to prevent potential attacks or server rundown.

Cost-Effective

Operating on LiteSpeed could be a low-maintenance task, especially in terms of support costs. The platform provides a greatly optimized support function and licensing costs are quite low compared to any other software upgrades.

With the range of listed features and the ease with which it can be optimized, LiteSpeed Web Server could serve as a perfect solution for someone looking for a low-cost, high-stability web hosting solution.

Control Panel Plugins

Popular hosting panels cPanel (WHM), Plesk, and DirectAdmin that enable administering LiteSpeed functions directly from the panel. This could be used to address functions like LiteSpeed installation, switching to and from Apache and restarting.

Virtual Host Templates and Web Admin Console

This feature offers users easy maintenance of multiple virtual hosts and can help to customize basic setups separately. There are also usage-configurable templates for uses such as W3 Total Cache or reverse proxy.

Per-IP Throttling and SSL Renegotiation Protection

LiteSpeed is equipped with customizable security features such as per-IP connection, bandwidth throttling, and request. This blocks IPs that have too many connections and prevent cyberattacks before they penetrate the server.

Brute Force Protection for wp-login.php

The WordPress protection system prevents attackers from damaging the server by attempting login to a target WordPress account using different passwords or passphrases. Wp-login.php and xmlrc.php are typically the main targets for such attacks.

Reduced Hardware Costs

The web server can reduce costs for repair and offers 24/7 support. 24/7 support can be contacted through the software and website for immediate action. Additionally, the software's integration to cPanel will give users access to the application's support team.

APACHE

Apache HTTP Server, mostly known as Apache is a free and open-source web server that distributes web content through the Internet. And since it is open-source software, it has made it very popular among developers trying to configure their own code in collaboration with Apache original source code. The web server has been around since 1995 and was one of the key enterprises that are considered responsible for the initial technological project and the growth of the Internet that it has promoted.

Apache has a great modular functionality at its core and can be adjusted do to whatever you want. It also has the ability to operate with large amounts of traffic and remove unwanted modules to run in a lightweight and efficient manner.

Some of the most popular modules that can be added to handle traffic are Server-Side Programming Support (PHP), and Load Balancing configurations. Apache can also be set up on Linux, Windows, and macOS. And if you know how to manage Apache on Linux, you can apply it on both Windows and Mac as well.

Apache main features include

- Auto-indexing

- Supports HTTP/2

- Gzip compression and decompression

- Bandwidth throttling

- Loadable dynamic modules

- Load balancing

- Session tracking

- URL rewriting

- Geolocation based on IP address

- Handling of static files

However, Apache is just one component that is needed in a web application set to deliver web content. There are three most common web application stacks are known as LAMP, MAMP, and WAMP.

- LAMP is an application stack in which "L" stands for Linux, "A" for Apache, "M" for MySQL database, and "P" goes for programming languages such as PHP, Python, or Perl. It is a light-weighted local server solely supported by and runs on the Linux OS and cannot be run on any other OS.

- MAMP is another local server, which supports (M) macOS and other web projects based on (A) Apache Server, (M) MySQL database, and (P) PHP programming language. It is mostly used for macOS but also provides all the equipment needed to run WordPress on the system.

- WAMP is a software package that includes Apache Server (A), MySQL database (M), and PHP script-based language (P). The "W" in WAMP stands to denote its exclusive use for the Windows OS. WAMP is used in Windows-based 32-bit and 64-bit systems and developed with PHP.

Apache functions as a way to communicate over networks from client to server using a wide variety of protocols. But the protocol that Apache is most known for is HTTP/S or HyperText Transfer Protocol (S stands for Secure).

HTTP/S is used to determine how messages are formatted and then transferred across the web, using instructions for browsers and servers that explain how various requests and commands should be responded to. Apache can as well accept and route traffic to certain ports and domains based on specific address-port combination demands. Normally, it runs on port 80, but

Apache can also be bound to different ports for different domains, allowing other websites and domains to be hosted in a single server. When a message reaches its destination or recipient, it sends a notice to the original sender acknowledging that the data has successfully arrived. If the server faces an error in receiving data, the destination host sends a Not Acknowledged message to inform the sender that the data needs to be forwarded again.

Two main Apache versions include:

- **Version 1.1:** It was approved for usage in 2000 with the primary change in derived products that are no longer required to include attribution in their advertising materials, only in their documentation.

- **Version 2.0:** Was adopted via Apache License 2.0 in January 2004. The stated goals of the license included improving compatibility with General Public License-based software, allowing the license to be included by reference, clarifying the license on contributions, and requiring a patent license. Additionally, version 2.0 enhanced the following features:[4]

  - **Unix threading:** Apache httpd can now run in a hybrid multi-process, multithreaded mode.

  - **New build system:** the build system has been rewritten to be based on autoconf and libtool. This makes Apache httpd's configuration system more similar to that of other packages.

  - **Multiprotocol support:** Apache HTTP Server now has some of the infrastructure in place to support serving multiple protocols.

  - **Multilanguage error responses:** error response messages to the browser are now provided in several languages, using SSI documents. They may be customized by the administrator to achieve a consistent look and feel.

  - **Simplified configuration:** many confusing directives have been simplified.

  - **Regular expression library updated:** Apache now includes the Perl Compatible Regular Expression Library.

---

[4] https://httpd.apache.org/docs/2.4/new_features_2_0.html, Apache

The fact that Apache HTTP has one of the most easily customized infrastructures, that it is reliable and secure makes it a common, by default choice for many top companies in the world. It is also true that every developer should learn to administer and configure Apache out of other servers first. And while the number of new web servers is increasing, Apache still plays a pivotal role as the backbone of the early Internet year.

You can tune your Apache server performance following this list of recommendations:

1. **Select the right MPMs:** As mentioned before, Apache is a modular server, meaning you can add and remove features easily. Multi-Processing Modules (MPMs) used for managing network connections and binding ports provide this modular functionality at the core of Apache. These modules let you add threads and could possibly even move Apache to a different OS. With that, only one MPM can be active at a time, and the traditional model of one process per request is called prefork. A newer, updated module is called worker, which operates multiple processes; and event MPM is a module that keeps separate strains of threads for different tasks. If you want to see which MPM you're currently using, try httpd-V, or httpd-l to see the list of compiled modules.

    Choosing what MPM to use depends on many factors. Prefork being the default MPM is claimed to be a safer choice. On the other hand, Worker MPM utilizes multiple child processes that can have multiple threads each, with each one of the threads handling one connection at a time. Worker usually preferred because of its high-traffic running capabilities as oppose to other modules. Event MPM is threaded in the same way as Worker MPM, but is created to allow more requests to be served simultaneously. It frees the main thread to proceed with other requests by delegating some converting work to supporting threads. And even though it functions identically to worker MPM, Apache calculates the lowest resource requirements when used under the Event MPM.

    Nevertheless, no matter what MPM you choose, it should be configured appropriately to be able to swiftly control the active and inactive workers and transfer threads or processes.

2. **Remove unused modules:** After you take a look at what Apache modules have been compiled by entering httpd-l, you can go ahead and disable unneeded modules from loading, including mod_php, mod_ruby, mod_perl, and others.

3. **Set extended status off:** ExtendedStatus is used to set off several systems calls for each request to gather statistics, which may impact the speed or server performance. Therefore, when possible it is advised to use ExtendedStatus only for a specific time period in order to complete analytics, but then turn it off.

4. **Disable hostname lookups:** Every time a web server receives an HTTP request, it saves that information in a log, while translating the user's IP address into a domain name. This DNS conversion costs additional time and should therefore be disabled.

5. **Utilize mod_gzip/mod_deflate:** Gziping your content before sending it to the recipient that has to uncompress on his end is going to minimize the size of file transfers and will consequently improve the overall user experience.

6. **Use mod_disk_cache not mod_mem_cache:** To avoid high memory usage use mod_disk_cache with a flat distribution. Set values CacheDirLength=2 and CacheDirLevels=1 to ensure htcacheclean does not take a lot of time when cleaning your cache directory.

7. **Move cache to an external drive:** To avoid slowing down server access, it is recommended to locate your cache on a separate physical disk like a Solid State Drive.

8. **Testing:** It is vitally important to perform a configuration test before and after tuning Apache. Testing can provide necessary information about the condition of files and potential syntax errors.

OPERATING SYSTEMS

A server OS could be described as "an interface between a computer user and computer hardware. OS is a software which performs all the basic tasks like file management, memory management, process management, handling input and output, and controlling peripheral devices such as disk drives and printers."[5] Mostly called server OS, OSs function within a client/server architecture, representing a software layer on top of other software programs or applications. The two most popular server OSs include Linux and Windows Server.

Linux

Linux is one of the many OSs that are considered to descend from the Unix multiuser computer OSs. Back in the late 1980s in the University of Helsinki, Finland, a small Unix-like OS called MINIX was created to demonstrate the principles of OSs to students. And since the MINIX code was available for educational applications, a 21-year-old computer science student named Linus Torvalds has written a new code inspired by MINIX that he then released as open-source under the General Public License. He combined the name MINIX with his own name and created the new OS Linux. Nowadays several OSs for smart devices, such as smartphones, tablet computers, home automation, smart TVs (Samsung and LG Smart TVs use Tizen and WebOS, respectively), and in-vehicle infotainment systems, are based on Linux. Major platforms for such systems include Android, Firefox OS, Mer, and Tizen.[6]

[5] https://www.tutorialspoint.com/operating_system/os_overview.htm#:~:text=An%20Operating%20System%20(OS)%20is,as%20disk%20drives%20and%20printers., Tutorialspoint

[6] https://techcrunch.com/2013/08/07/android-nears-80-market-share-in-global-smartphone-shipments-as-ios-and-blackberry-share-slides-per-idc/, Techcrunch

The core of the OS in Linux is the kernel. It is also the very thing that makes Linux stand out from other OS, even from MINIX. And even though Linus Torvalds was once accused of stealing MINIX code, you can presume the opposite just by glancing at two kernels. MINIX is based on a micro-kernel that holds the bare minimum amount of code needed to run an OS, while Linux has a monolithic kernel that manages most of the processes like system calls, filing systems, and even virtual memory. Another Linux kernel advantage is that while the internal features of the OS can progress over time, the interface between the kernel and user applications space remains anchored, which means if you upgrade Linux, you will not be forced to upgrade your programs.

The user interface, also known as the shell, is either a command-line interface (CLI), a graphical user interface (GUI), or controls attached to the associated hardware, which is common for embedded systems. For desktop systems, the default user interface is usually graphical, although the CLI is commonly available through terminal emulator windows or on a separate virtual console.[7] As mentioned, Linux's code is open-source that is released under the General Public License. This permits users to:

- Download and install the Linux OS for free

- Enjoy unlimited redistribution privileges of all the original and modified versions for free as well

- As an administrator, you can study and customize the OS

Linux has also been promoted as a community-based project, meaning that the kernel is maintained not incorporate but rather in open access via a public email list, where all the proposed updates and discussions on whether they should be implemented take place. Anyone can join the Linux community by contributing to this mailing list.

Such a great communal spirit of Linux also means that there is always a huge group of experts as well as enthusiasts participating in a quest to advance the system. Due to this factor, it now has an agile ecosystem with a networking stack optimized for modernized databases. Additionally, Linux is designed with separate control and data-forwarding planes that offer a number of powerful capabilities for system administration. And

---

[7] https://en.wikipedia.org/wiki/Linux, last edited on December 9, 2021

while many programs are already included with most installations of the Linux OS, you might want to add more to reach the functionality level you need. Normally for that, you will have to access a package manager, a program that downloads software sets from available for free software libraries.

Most of the networking applications for Linux, especially those focusing on traffic analysis, security, and network management have to be regularly analyzed and consequently optimized using one of the multi-faceted optimization techniques:

The CPU

The CPU is the fundamental component of the OS. The level of speed and overall performance of the CPU largely affect the capabilities of the whole OS. Thus, ensuring higher CPU frequency will certainly result in better server performance. For instance, you may consider adding a hyper-threaded processor to the CPU so it can run multiple threads at a time instead of only one thread. For such modulation, the configuration and performance of the CPU should be placed at the main location.

Random-Access Memory (RAM)

The size of RAM is another important factor affecting Linux performance. If the memory is too small, system processes will be disjointed and slow, which may even result in loss of response; in case if the memory is too large, you might be wasting a lot of resources. Because Linux uses physical and virtual memories, virtual memory can help avoid the potential shortage of physical memory. But in order to ensure the balanced, high-performance operation of the application, it is generally recommended to install a 64-bit OS and enable Linux's large memory kernel support. Under the 64-bit OS, RAM can meet the memory usage requirements of all applications. Features that may experience memory performance limitations include cache servers, NoSQL servers, and database servers; to prevent any obstacle for such applications, the memory size should be placed at the primary location.

System Installation Optimization

When installing a Linux system, the partitioning memory stores and disks can affect the running performance of the system in the future. Therefore, it is advised to start system optimization right from the installation of the

OS. The following allocation of the Redundant Array of Inexpensive Disks (RAID) data storage virtualization technology can follow almost all the requirements of the OS applications.

RAID 0 would be suited for applications with low-security requirements; and those with high data security and no limitations for reading and writing, the disk can be made into RAID 1. And for applications with high read and write demands and specific security requirements, RAID 10/01 can be picked. RAID 5 can fit applications that do not require special write operations data security. This is the common optimization method of pairing different RAID levels together with different application requirements.

File System Optimization

Another part that plays an important role in system resource optimization is file system optimization. The Linux standard file system starts from VFS and then ext, and the optional file systems under Linux include ext2, ext3, ReiserFS, ext4, and xfs. Different file systems are selected according to different application requirements. In order to advance the log file system, the XFS high-performance 64-bit journaling file system could be installed. XFS ensures high-bandwidth access to data by distributing disk requests, low-latency, cache consistency, and fast data location. Therefore, it is recommended to test-drive XFS as an excellent tool to accelerate your logging capabilities and fast write performance.

Kernel Parameter Optimization

After the file system is installed, the optimization work could be switched to the system kernel parameter. Nonetheless, the optimization of the kernel parameters should be contemplated together with the applications deployed in the system. Meaning if you are deploying a web application, you need to optimize the network parameters according to the characteristics of the network kernel parameters such as net.ipv4.ip_local_port_range, net.ipv4.tcp_tw_reuse, net.core.somaxconn. And if, for instance, the system is deployed with an Oracle database application, then you would have to share the memory segment (kernel.shmmax, kernel.shmmni, kernel.shmall), filehandle (fs.file-max).) and system semaphore (kernel.sem), together with other parameters to optimize settings.

Windows

Windows computer OS was initially developed by Microsoft Corporation to run personal computers. The first-ever version of Windows was released in 1985, featuring the first GUI as an extension of Microsoft's existing disk OS. It was the first OS to allow IBM-compatible PC users to visually navigate a virtual desktop, be able to open graphical "windows" with contents of electronic folders and files with the click of a mouse button. Later versions have developed a more multi-functioning interface adding Windows File Manager, Program Manager, and Print Manager programs. And when the 1995 Windows has introduced built-in Internet support, including Web browser Internet Explorer, the official domination of the PC market has been established.

With the 2001 release of Windows XP, Microsoft has abandoned the Windows 95 kernel for a more high-powered code base and offered more practical memory management. With that, it has been able to put forward new Windows packages that provide multiple editions for multimedia developers, businesses, and consumers. The next version of Windows Vista that came along in late 2006 has met a great deal of marketplace resistance, getting a reputation for being a slow and resource-intensive system. Quickly after that, in response to Vista's disappointing review rate, Microsoft released Windows 7. Unlike Vista, a new edition had modest system demands and great speed improvements.

Windows 8 in 2012 presented the synchronized settings feature that could help to log on to another Windows 8 as well as offered a new screen display that set OS applications appearing as colorful tiles. Windows 10 was released in 2015, replacing the Internet Explorer web browser with Microsoft Edge and introducing a digital personal assistant Cortana. It was also announced that Windows 10 would be the last version of Windows, meaning that users would only receive regular updates to the OS without any wide-ranging revisions to come.

With that, Windows also runs additional versions such as Windows CE (officially known as Windows Embedded Compact), for on minimalistic computers, like satellite navigation systems and some mobile phones. And Xbox OS is an unofficial name given to the version of Windows that runs on the Xbox One. This version has a more specific implementation with an emphasis on virtualization as it is three OSs running at once, consisting of the core OS, a second implemented for games, and a more Windows-like environment for applications.[8]

We have combined a list of Windows best features that make it so popular up to this day.

Speed

One of the most straightforward characteristics of Windows OS is speed. Aside from incompatibilities that many users faced with Vista, other versions from 7 to 10 have a more responsive and rapid reaction to it. Microsoft has also recognized the need for improved desktop receptiveness and has spent a lot of time and effort getting the Start Menu service just right.

Later with Windows 10, you can also choose between different performance plans. Thus, in order to boost your PC performance, you can change your power plan from "Power saver" to "High performance" or "Balanced" to get an instant performance boost.

Search and Organization

Microsoft has introduced the concept of Libraries, advancing "My Documents" concept to another level. You can now match various Libraries, such as Documents and Pictures, with multiple locations that you add and delete yourself, taking control of your own database. Another thing is that is worth mentioning is the Windows 7 improved search tool that enables you to effortlessly navigate around the Desktop.

---

[8] https://www.extremetech.com/gaming/156467-xbox-one-hardware-and-software-specs-detailed-and-analyzed, Xbox

Safety and Security

Staple security features in Windows include two new authentication methods by PINs and picture passwords. In addition to that, the Microsoft team has added antivirus settings to Windows Defender Smart Screen and the Secure Boot functionality on unified extensible firmware interface systems to protect the boot process against malware. Windows 8 version also provides incorporated system recovery that comes with "Refresh" and "Reset" functions, including system recovery from USB drive. In regards to family safety, Windows has a great Parental control function, which allows parents to monitor children's activities on a device by getting regular activity reports and safety controls.

Interface and Desktop

Many of the Windows currently ongoing updates are related to introducing significant changes to the OS' interface that users mostly experience on tablet computers and other touchscreen devices. The new user interface is based on Microsoft's Metro design language and uses a Start screen format that was previously used for Windows Phone applications. The Start screen was initially created to display a customizable array of live tiles linking to various web applications that could be rearranged on a desktop in accordance with users' preferences. And due to the trend of multitasking, a new touch-optimized settings application known as "PC Settings" was introduced. It is mostly used for basic configuration and does not include many of the advanced options that could be accessed from the normal Control Panel.

Taskbar Menu

At first glance, the taskbar looks like it has not changed much since Vista. However, that is not the case, and many of the new features could be visible at the first glance. Thus, starting from Vista taskbar icons are now made larger, and items are grouped together without any labeling. If you have multiple Word documents or Windows Explorer windows, try opening them, and you will see a stack appear on the taskbar. If you want to check any single content, you can just hover the mouse over the app and each window or document will be visible in a thumbnail. And in case you need to see the list of recent documents you can point on the Start menu, a small arrow to the right of applications will expand and give all the information which then can be pinned to keep the file permanently on the list.

To conclude, the biggest advantage of Windows is that it offers instant solutions that can be implemented by anyone who has ever owned a computer. It is as well compatible with any file or document produced in the office space.

Probably the only other major disadvantage of using Windows is the fact that over 95% of all viruses and malicious software are scripted for the Windows OS. This means that you need to watch your security measures and regularly check analytics reports if you are a Microsoft software user. Another thing many developers mention is Windows needs more stale software services in comparison with other Mac or Linux counterparts.

We see no need to indulge in the dilemma of deciding which OS to use when configuring your module performance. As it would be wrong to assume that OS the computer runs on the need to have a standard set of management tools. Today, most administrators can remotely operate servers, therefore making the type of OS your computer manages seems irrelevant. However, if you are going through a decision-making process, it is recommended to base the calculus on other primal things like the cost of OS, added fares of support service, and technical competence of your team to run on either Linux or Windows.

PHP OPTIMIZATION

PHP is a scripting language initially invented by Rasmus Lerdorf in 1995 for personal use. At first, "PHP" originally stood for "Personal Home Page," but later when Lerdorf expanded its capabilities, PHP came to stand for the acronym "PHP: Hypertext Preprocessor."

PHP runs on different OSs, like Linux and Windows, and supports various databases such as MySQL, Microsoft Access, and Oracle. PHP is considered as multifunctional as it can not only collect data, but it can also read, write, delete, and close files on the server. Additionally, it can be easily be embedded in HTML with tags <?PHP? >. The scripting language is mostly used for server-side scripting, command-line scripting, and desktop applications. For the latter, it provides a great object-oriented user interface and lets you include images, PDFs, videos, and sounds into your HTML.

Before starting to design a PHP application, it is advisable to run benchmarks and adequately test data on your hardware and software to condition your performance parameters. This analysis is important to help and guide your coding knowing all the risks and potential trade-offs. The latest version of PHP 7.4 (active support until November 28, 2021) is one of the fastest, if not the fastest version of PHP out there. It has no compatibility issues when migrating between other versions and by far outweighs all the previous editions in development cost and time for the modifications. It is therefore recommended to upgrade to a current version for better PHP performance in case you are using older versions.

If you are developing a web application based on PHP, you should take into consideration server resources that contribute to the speed and price of the hosting like storage, memory, and number of CPUs. Optimizing these parameters in accordance to project requirements will allow the application to run smoothly. Some programmers tend to move the fine-tuning of code for the end of a project cycle. However, if you do not want to end up rewriting parts of code in the end, it is only advisable to schedule tests during the development process. There are many ways to handle PHP performance optimization in web applications and the following are just a few of them:

Single and Double Quote Usage Matters

You may start small and think about the way to improve using double and single quotes. It is a well-known fact that the usage of single quotes for strings matters. It has been proven that single ones, especially in larger loops and strings, run significantly faster than double-quotes. Double-quoted strings usually look for some variables before displaying the string itself, which tends to slow it down. To illustrate:

```php
function doubleQuotes($iterations){
    $temp_str = "";
    $start_time = microtime(true);

    for ($x=0; $x<$iterations; $x++) {
        $temp_str. = "Hello World! ";
    }

    echo "doubleQuotes(): ". (microtime(true)-$start_
time).  "</br>";
}

function singleQuotes($iterations) {
    $temp_str = ';
    $start_time = microtime(true);

    for ($x=0; $x<$iterations; $x++) {
        $temp_str. = 'Hello World! ';
    }

    echo 'singleQuotes(): '. (microtime(true)-$start_
time).  '</br>';
}

doubleQuotes(500000);
singleQuotes(500000);

doubleQuotes(): 0.065473079681396
singleQuotes(): 0.027308940887451⁹
```

You can see from this demonstration that the string that contains single quotes runs more than twice as fast in comparison to the string test with double-quotes. The small distinction in milliseconds might look insignificant, but in terms of performance optimization, it would help PHP web applications to gain an extra edge and enable hundreds of users to access it error-free every minute.

Close or Unset Variables

In order to quire the database, you should a connection has to be made through declaring a connection variable. To optimize this practice, it is

---

recommended to close the connection after the query or all the queries are complete. With that, similar to the connection variable, after reading or writing to the file, that link must be closed as well. Ending connections will considerably help to save memory usage, especially at times when multiple people are accessing the same request of a web application.

```
$conn = new mysqli($servername, $username, $password,
$dbname);
//queries here
$conn->close();

$myfile = fopen("sample-file.txt", "r") or die("Unable
to open file!");
//read the contents
fclose($myfile);
```

Optimize SQL Queries

It is viewed as misconduct to have an HTTP request that has multiple database queries. If related database tables cannot be queried by applying joins, then it has to be reassigned. Another way to optimize performance with SQL queries is to add indexes to some columns. And even though indexes require additional storage space, unlike just regular columns, having an improved retrieval rate affects user experience. This way, the retrieval of records through indexed columns will get faster. Usually, the columns that need to be indexed are those used in join, group by, order by, and where clauses.

Minify Your CSS and JavaScript

Another effective way for PHP performance optimization is minifying JavaScript (JS) and CSS code. It not only improves load times by reducing your code in size but also enables browsers to quickly parse these files because comments and white spaces are left out of the way. This procedure will make the code unreadable by humans, but the readability of code should not be a priority of web applications anyway.

Use a CDN

The best way to load web application libraries is through the content delivery network (CDN). Especially those applications that need PHP performance boost must consider using CDNs to download resources. Typical

CDN allows users to download contents from a closer source, rather than loading it from where the entire application is hosted, helping to reduce the loading time of the application. Also, most of the usually utilized images, CSS or JS files are static anyway, so it is only right to keep the cached copies of the content at a server closer to where the users are. This way, data travel a shorter distance and will be displayed faster, diminishing latency in your application.

Monitor PHP Performance

If you want a steady performance level for your PHP applications, make sure you invest in a suitable monitoring tool. That way you can keep track and trace every web request in your application and generally run your applications faster than usual. One measure to focus on would be the PHP elements loading times. Monitoring this parameter will give you a chance to see which part of the application is underperforming and needs to be tuned. Another thing to consider is the traffic and overall responsiveness of your application. You want to check whether your server has the capacity to handle requests and respond to specific PHP modifications while giving access to a great number of users.

In other cases, PHP performance errors happen during database queries and API calls because the application has to wait for these processes to complete before moving on to the next task.

To conclude, those were the six easy-fix ways to improve the quality of your PHP applications. The process of PHP application performance monitoring involves collecting detailed information about your application services, transactions, database queries, and other performance metrics, so you can identify and resolve issues before they impact end users. Applying one of the mentioned optimization methods on their own is expected to deliver a notable performance improvement. Nevertheless, when implemented together, you are guaranteed to see significant performance and quality improvement.

Conclusion

By this point, we hope you find yourself comfortable with the fundamentals of web development and have the ability to administer and balance at least those optimization methods covered within in order to provide a positive, operative experience for your audience. However, each topic that we have discussed was just the tip of the iceberg. In reality, the techniques and approaches we covered go into much greater depth, meaning you still have far more to learn.

This book took great effort to introduce you to the web world, but in many instances, each of these chapters should be addressed by different parties. It is a common practice that most companies would typically have one or more employees focusing on database development issues, server-side optimization, etc. Web analytics and front-end design alone have shaped entire divisions. How you create your team should not only be determined by the size of your project or your finances but also by the depth and complexities you might be willing to face with it.

We have established that improving your website performance starts from day one. And since your businesses' marketing and operation plans might constantly be expanding and evolving, committing to better user experience and Search Engine Optimization (SEO) would require consistency and a systemic approach.

True optimization is an almost daily duty that involves monitoring changes in rates, rebranding, and redefining the site as content or design changes. With that, there still might be some things that you cannot control like outside links or redirected traffic. And if you find yourself confused, it is good to keep these few basic tasks in mind:

- Make sure you are mobile optimized
- Update your content regularly

DOI: 10.1201/9781003203735-8

- Use meta tags and tags to emphasize important content

- Do not include too many keywords in your meta tags

- Integrate with social media for wider exposure

By mastering these simple steps of website optimization, you can expect to put your product on the map and enable heavier and more organic traffic. Another area to concentrate focus on would be technical optimization. And we went through standard modules that you may implement to raise off-page efficiency. Those methods included monitoring sitemap errors, console errors, keeping your code and URL clean, and ensuring good crawlability. You can improve crawlability by regularly administering your site structure (HTTP aspects, metadata) making sure bots have clear access to your content.

Smart front-end optimization and front-end testing are another way for developers and web owners to analyze and examine site performance from a user's perspective. It has become necessary to test how technical processes shape front-end perception and what can be done to prevent any errors from happening. Dismissing front-end performance testing might harm businesses in terms of maintaining high operation standards and customer retention. The back-end and front-end features need to work hand in hand as there would be no traffic if your back engines are malfunctioning; yet even technically mega-advanced site will not attract the needed conversion rate if the content is irrelevant or dull. Content optimization, therefore, should be treated as an integral part of web development because it is responsible for optimizing elements that bring style and interest like image, video, and news content.

In order to serve your target audience with materials they would like, you need to understand who your viewers are, which you can find in your website logs. You can learn a great number of things with analytics which can later result in new website features and tailoring to popular interests or recent searches. Reviewing analytic reports, you might find out information about your visitors that you had not anticipated—that may inspire you to add additional language support, try to use mobile devices, or redesign your site to provide more content or a smaller interface. Tracking every action that your user takes can provide insight. Collecting this data across a volume of users over time will certainly give you the means to discover things you were not even aware of.

Web analytics is a broad domain that you started to navigate by being introduced to a huge range of concepts and terms such as unique visitors, page views, bounce rates, and conversion rates. Having correct analytics is really important for understanding what needs to be optimized in terms of your localization efforts as well. It allows you.

Nowadays it is impossible to do business without considering social media. Access to social media platforms will help you to establish an actual customer journey around your site and obtain more insights and actual feedback about design and content. That's why it is important to see not only the headline figures on how many times your content has been shared, it is also necessary to assess interaction, engagement, conversation, reach, and others. When considering social media analytics, you should be looking at major indicators of potential reach such as the number of followers, level of engagement, amount of traffic, and reputation (social media channels are a great platform to find out how others perceive your business and respond accordingly). The actual data files will not be easy to interpret and seek answers from unless you are willing to spend man-hours working with them. Instead, most people who interact with logs prefer to use an outside program that reads the files and helps them see trends over time so they can make more use of the raw data itself. For that, you can choose from a variety of free, open-source solutions that can compile your data into graphs and tables focusing on aspects that are more important for you at the moment—where visitors are coming from, what browser or device they used, what pages are more/less popular.

When you are ready to start looking for a suitable online analytics solution, it is important that you engage in scrupulous and extensive research before committing to any tool. Some of them could be harder to install while maintaining others may demand special skills from your team. The process of decision-making can be time-consuming and will require you to go through standard demands for data analytics and legal and procurement-related purchasing constraints. We recommend approaching this managerial situation with a team mindset. Deciding what analytics solution to opt for should start with a thorough team brainstorming session. Ideally, everyone has to contribute and present professional opinions as to which platform would serve their objectives the best. A multi-functional pot that consists of experts from different fields such as marketing, engineering, support, and research departments has to be created.

And since website analytics can strongly reinforce both qualitative research and quantitative results when reviewing your KPIs, you want it to comply with the best practices of analysis provision:

- **Avoiding only statistical reports:** Reporting only large numbers could be deceiving and does not provide a full context of site performance. You want to see it in-depth and learn about what is behind the data provided and what needs to be done to attract more conversion rates.

- **Always provide data with insights:** In addition to the previous point, every report should be coupled with some meaningful information regarding how the data shows areas of success and failure.

- **Encouraging data-driven decision-making:** The website analytics tool you opt for should tell you whether you have met or failed to meet your goals and help you work on how to improve your KPIs. Sometimes it would be necessary to directly involve the tech team in the decision-making process of improving web traffic and conversion flows.

- **Avoiding being snapshot-focused:** Paying attention to all sorts of metrics can provide a longer-term understanding of users and will allow you to evaluate how the website evolves over time.

Getting access to website performance data flow should enable you to better prioritize your assignments and efforts in marketing and SEO campaigns. But in order to effectively capitalize a set of metrics, one has to have a means of actually acting on the data being presented. Being able to review and react accordingly to the critical directions is essential to creating successful strategies.

To store provided analytics and other generic information, many sites create databases to help them track everything about their site, customer preferences, and overall experience. This material will later be applied to regulate larger questions like how best to get the customer to return or what SEO aspects should be modified. Depending on who uses the database and for what purpose, it can be either a static or dynamic solution. For any entity that has even moderate structure and data complexity like e-commerce sites, social networks, or blogs, it is important to switch to a database management system to preserve and operate files in an error-free manner. At the moment, there are three commonly used database management systems out there: widely popular MySQL, PostgreSQL that has gained audience due to advanced Indexing and multi-version concurrency control, and MongoDB, which is the biggest open-source non-relational data management system.

In the final part of this book, we have decided to cover the top three web server software that have different abilities to handle server-side programming, site-building tools, and security characteristics:

- **Nginx:** A popular open-source web server that is known for its unique event-driven architecture.

- **Litespeed:** A free web server is seen as one of the fastest due to lesser central processing unit power consumption.

- **Apache HTTP Server:** An open-source web server is used for Windows, Mac OS X, Unix, Linux, Solaris, and other operating systems.

These three vary in configurations and set of default values. And when considering which high-performance webserver to opt for, we advised examining how well it works with the operating system. The two most popular servers that were mentioned included Linux and Windows Server. Both function within a client/server architecture, representing a software layer on top of other software programs or applications. The scripting language is used for server-side scripting, command-line scripting, and desktop applications in Linux and Windows is PHP. PHP also supports various databases such as MySQL, Microsoft Access, and Oracle. The programming language is considered multifunctional as it can not only collect data, but it can also read, write, delete, and close files on the server. Additionally, it provides a great object-oriented user interface and lets you include images, PDFs, videos, and sounds into your HTML.

Once you are ready to move beyond programming alone and into multiple teams working on the same project, a management approach will be required to determine pace, goals, deadlines, and to maintain order and understanding of the project. Nowadays, there are a lot of management tools available, but one of the most widely-used solutions is Agile.

Ultimately, the guiding principle of Agile is to work on creating a mindset and atmosphere where you get the best people by trusting them and focus more on the customer's wishes and the project objectives more than on micromanaging internal processes and bureaucracy.

We find it extremely useful to keep in mind the following 12 principles of Agile development, according to a published manifesto by seventeen software developers:[1]

1. Our highest priority is to satisfy the customer through early and continuous delivery of valuable software.

2. Welcome changing requirements, even late in development. Agile processes harness change for the customer's competitive advantage.

3. Deliver working software frequently, from a couple of weeks to a couple of months, with a preference to the shorter timescale.

4. Business people and developers must work together daily throughout the project.

5. Build projects around motivated individuals. Give them the environment and support they need, and trust them to get the job done.

6. The most efficient and effective method of conveying information to and within a development team is face-to-face conversation.

7. Working software is the primary measure of progress.

8. Agile processes promote sustainable development. The sponsors, developers, and users should be able to maintain a constant pace indefinitely.

9. Continuous attention to technical excellence and good design enhances agility.

10. Simplicity—the art of maximizing the amount of work not done—is essential.

11. The best architectures, requirements, and designs emerge from self-organizing teams.

---

[1] http://agilemanifesto.org/, Agile manifesto

12. At regular intervals, the team reflects on how to become more effective, then tunes and adjusts its behavior accordingly.

Having learned this information should make web testing and optimization easy for you. Now it is time to get some real projects completed. After a few full implementations, you will learn more facets and details of one or more of the topics we covered. Once you succeed in every area of web practice, you are guaranteed to experience a more transformative impact than it will cause. You can start by creating your optimization plan using our guidelines and tips to avoid common mistakes and get insights into how you can lead this process. We hope that this book will serve as a treasure trove of excellent resources to go deeper into any of the areas explored here.

Index

Printed in the United States
by Baker & Taylor Publisher Services